Quality of Experience for Multimedia

FOCUS SERIES

Series Editor Abdelhamid Mellouk

Quality of Experience for Multimedia

Application to Content Delivery Network Architecture

Abdelhamid Mellouk
Said Hoceini
Hai Anh Tran

iSTE

WILEY

First published 2013 in Great Britain and the United States by ISTE Ltd and John Wiley & Sons, Inc.

ISTE Ltd
27-37 St George's Road
London SW19 4EU
UK

www.iste.co.uk

John Wiley & Sons, Inc.
111 River Street
Hoboken, NJ 07030
USA

www.wiley.com

Library of Congress Control Number: 2013946297

British Library Cataloguing-in-Publication Data
A CIP record for this book is available from the British Library
ISSN: 2051-2481 (Print)
ISSN: 2051-249X (Online)
ISBN: 978-1-84821-563-4

MIX
Paper from responsible sources
FSC® C013604

Printed and bound in Great Britain by CPI Group (UK) Ltd., Croydon, Surrey CR0 4YY

Contents

List of Figures

Preface

Owing to the emergence of different kinds of communication and networking technologies (core networks with intra-domain and inter-domain challenges, access networks, aggregation networks, spontaneous networks, the Internet of Things, etc.) and the current and foreseen proliferation of different and specific types of services supported by these technologies (real-time services, IPTV, VoD, social networking, E-Health, multimedia, gaming, smart cities, etc.), traditional network control strategies are not sufficient for handling the complexity and diversity of information in use that we see today. There is a strong need to develop a new paradigm to obtain the continuity of network services based on the new concept of smart communications and user's interaction. User perceptions, such as emotional, intense or satisfactional, is a natural way to interact with the real world, and will serve as a powerful metaphor for interacting with online services and performing an affective computing paradigm. In addition, the use of artificial intelligence tools together with biologically inspired techniques is needed to control network behavior in real time so as to provide users with the quality of service that they request, and to improve network robustness and resilience based on continuous user feedback. The key idea of this book is to present a new paradigm driven by user perception and, based on control theory and machine learning techniques in order to support smart communications, to avoid any complete interruptions in the whole chain of service treatment. The book focuses on the idea of how to integrate the Quality of Experience (QoE) into a control/command chain in order to construct an adaptive network system. More precisely, in the context of content-oriented networks (CON) that are used to redesign the current Internet architecture in order to accommodate content-oriented applications and services, an end-to-end QoE model is applied to a content distribution network architecture with respect to the following steps:

– To develop at the theoretical level an appropriate unified formal model based on bio-inspired modeling and knowledge distribution in order to construct a scalable and robust environment with low complexity for large-scale dynamic networks.

– To study at the empirical level why and how user perception can be quantified, analyzed and influenced.

– To create at the conceptual level a general framework for protocol stacks dedicated to smart communications featuring multiple technologies, devices and users.

– Finally, to build at the engineering level the appropriate model of programming abstractions and software architecture to develop a full-scale framework.

To give a complete bibliography and a historical account of the research that led to the present form of the subject would have been impossible. It is thus inevitable that some topics have been treated in less detail than others. The choices made reflect, in part, personal taste and expertise, and in part a preference for very promising research and recent developments in the field of QoE-based technologies.

The authors hope the readers will enjoy reading this book and gain many helpful ideas and overviews for their own study.

Abdelhamid MELLOUK
Said HOCEINI
Hai Anh TRAN
September 2013

Introduction

In recent years, we have been witnessing a rapid development of interconnecting applications. In addition to those that contributed to the popularity of the early Internet (e-mail, file transfer, etc.), there are now applications that rely on more sensitive network data. They include sound applications (voice, music programs, etc.), image applications (television programs, videoconferencing, video on demand, etc.) and urgent information applications (market orders). Based on a convergence of network technologies, the next-generation network (NGN) is being deployed to carry high-quality video and voice data. However, it is important for the operators and providers not to lose sight of the reason behind this new infrastructure: to provide network service that users want to use. Realising this idea means assuring positive experience for end users. The operators always try to deliver network services in the most cost and resource efficient manner with ensured user perception. Therefore, service providers are switching their focus from traditional Quality of Service (QoS) to user satisfaction, which is the overall success of a network from the user's perspective. According to a well-known expression, "The only way to know how customers see your business is to look at it through their eyes". In fact, the convergence of network technologies has been driven by the converging needs of end users. The perceived end-to-end quality becomes one of the main requirements of users that must be guaranteed by the network operators and the Internet service providers, through manufacturers' equipment. This is referred to as the Quality of Experience (QoE) notion that becomes commonly used to represent user perception. The QoE is not a technical metric, but rather a concept consisting of all elements of a user's perception of the network services. To ensure good end user perception, the service providers have encountered challenging QoE issues, such as QoE monitoring, measurement, diagnosis and management. Therefore, all of the network functions, such as admission control, access network selection, routing, resource allocation, transmission control and server selection, are expected to be adaptive to users' QoE. Of these fuctions, we focus on the routing and server selection.

In the past, the network operators used QoS notion as a method for improving their ability to accommodate data flows and to ensure QoS guarantees from the network. We take an example of a routing issue to show existing problems. The QoS routing allows the network operator to determine a path for every data flow through the network with sufficient resources to meet the flow's requirements. These requirements are based solely on QoS parameters, such as end-to-end delay, bandwidth and loss rate. Thus, identifying a path capable of guaranteeing these requirements needs the accuracy of the knowledge/information of the availability of network resources. In other words, a successful QoS routing depends on the accurate information of the available resources in the network. Most existing routing algorithms assume the availability of precise state information. However, in many cases, this information may be inaccurate and insufficient. The advertising updates in a dynamic network environment cause considerable imprecision for state information. So, the broad issue is the QoS routing with inaccurate state information [GUE 99].

The second issue related to the problem of complexity. To meet the requirement of a source node, apart from the resource parameters (e.g. end-to-end delay, loss rate and bandwidth), a QoS routing system may also try to optimize other criteria such as network utilization, carried load and number of flows successfully routed. These criteria can influence the routing process, but satisfying all of them is not easy. For example, the network provider may identify a path satisfying the requirement of the data flow, but cannot use it because it is too expensive in terms of the amount of consumed resources. Optimizing all of these criteria is known as an NP-complete problem. In reality, an end-to-end QoS model with more than two non-correlated criteria is NP-complete [WAN 96, MIC 79]. In order to resolve this problem, due to its complexity, network providers should integrate the QoE into their network systems.

In addition, a QoS network system addresses the different challenges associated with the accurate computation and estimation of network parameters. Previously, most projects/approaches were network oriented and only considered the service-level agreement (SLA) between network operators and service providers. However, this perspective is no longer sufficient with the new types of content on the Internet: the high-quality multimedia content. The latter has dramatically increased in the recent years. Users are downloading high-quality video streams, watching videoconferencing and broadcasting their own video streams on the Internet. This sensitive traffic in terms of end-to-end delay and loss rate, impose tighter constraints on the end users for a reliable content delivery, opposed to agreements of classic SLA. Thus, the interest of both network operators and service providers in providing satisfactional perception to the end users becomes more important than just considering technical network parameters.

Hence, the QoE has been introduced to resolve these above-mentioned problems. This new paradigm promotes the importance of the user in a whole system. With regard to QoE definitions, according to the International Telecommunication Union (ITU), the QoE is defined as the overall satisfaction of a service perceived subjectively by the end user. It is a measurement of how well a service satisfies the end user's expectations of the service and the complete end-to-end system effects. According to [HAS 10], QoE includes relevant factors such as a user's internal state (e.g. predispositions, expectations, needs, motivation and mood), the characteristics of the designed system (e.g. complexity, purpose, usability, functionality and relevance) and the context (or environment) within which the service is experienced (e.g. organizational/social setting, meaningfulness of the activity and voluntariness of use). Today, despite the meaningful progress made, QoE is always a significant interest and continuing challenge of academic and research industries.

The trend towards using QoE in network management systems is explained by the appearance of emerging multimedia services in the Internet, such as mobile online gaming, immersive environments, 3D virtual world, social networking and IPTV applications. The service quality depends not only on technical QoS parameters but also on end user perception. The challenge between network operators becomes satisfying end users who play an important role in the quality management model. In fact, the clients are not only "the spectator" but also the key "actor" of network management. The QoE has been introduced to face this new challenge.

Improving the quality of the service as perceived by the users is a significant challenge of network services. In this context, the goal is to minimize their customer churn and yet to maintain their competitive edge. Based on this kind of quality trade-off, QoE [SHA 10a, BRO 10, TRA 10] has been introduced as a combination of user perception, experience and expectations instead of using only technical parameters mainly gathered from QoS measurement. The QoE management takes into account the needs and the desires of the subscribers when using network services, while the concept of QoS only attempts to objectively measure the service delivered. As a critical measure of the end-to-end performances at the service level from the user's perspective, the QoE is a kind of metric for the system design and engineering processes that can only be measured dynamically at the very end of any transmission activity.

For these reasons, we focus on the idea of how to integrate the QoE into a control/command chain in order to construct an adaptive network system. More precisely, in the context of content-oriented networks (CON) that are used to redesign the current Internet architecture in order to accommodate content-oriented applications and services, we apply the end-to-end QoE model to a content distribution network (CDN) architecture that is composed mainly of two layers: the *routing* and the *meta-routing*. On the one hand, the routing layer takes into account the routing process required to route a data packet between clients and servers. In this layer, we proposed a routing algorithm based on user perception. On the other hand, the meta-routing layer includes various modules such as *server placement*, *cache organization* and *server selection*. We focus on the third module: the server selection, which selects the most appropriate replica server among a server set.

As mentioned previously, the QoE is today becoming a decisive criterion for network management success. In this book, we present a new adaptive control model to carry the QoE notion. Our idea is to design an adaptive loop for improving the service quality of a network system by taking into account the end user feedback. In our model (Figure I.1), the *QoE measurement module* evaluates the user feedback and sends the result to *the control module*. The latter analyzes the received results, makes a decision and sends it to *the command module* that will apply the chosen action to the network system. This loop model refers to an adaptive system based on the end-to-end QoE.

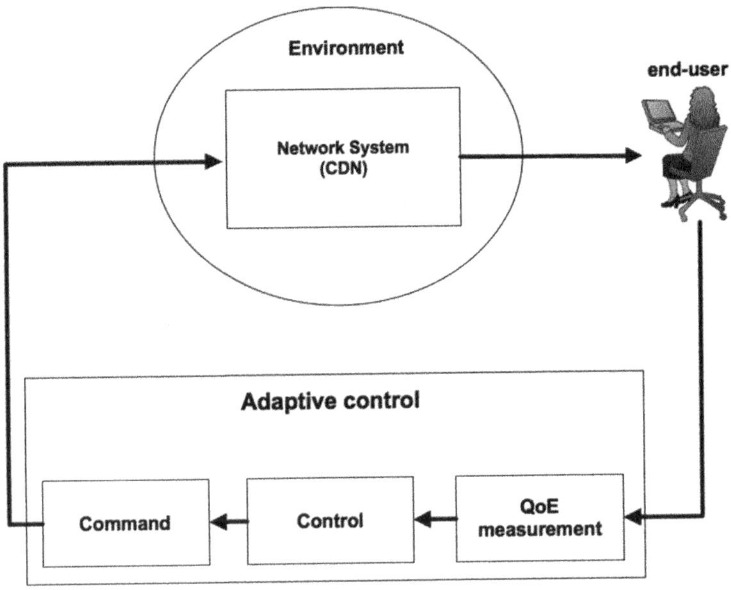

Figure I.1. *Adaptive control/command model*

To enhance the practical significance of the proposed model in Figure I.1, we applied the end-to-end QoE model above to a real network infrastructure: the CDN. The latter has been introduced to improve the delivery media content service. It is an effective solution to improve network quality and to optimize resource utilization. The main idea of CDN operation is to move the content from the origin server to the servers close to end users, namely replica servers. Based on the idea of applying the end-to-end QoE model to a CDN, we designed a QoE-based content distribution network architecture (QCDNA) whose goal is to maintain the best QoE in a dynamic network environment.

The book is organized as follows. In Chapter 1, we present our vision for bio-inspired control theory for a user-driven smart communications paradigm. We also talk about the user's perception metrics, affective computing and the knowledge dissemination.

In Chapter 2, we briefly survey the QoE notion. First, we talk about the QoE concept, what the QoE is and how important it is in the networking context. We also present some standard organizations studying QoE, such as ITU-T, Broadband-forum and TeleManagement-forum. Then, we talk about the relationship between the QoS, the widely used notion, and the QoE, the new view of network operators. In this chapter, the impact of networking on the QoE is presented.

In Chapter 3, as this book focuses on integrating the QoE metrics into a CDN architecture, we present the CDN by describing their two main layers: the routing and the meta-routing. For the routing layer, we describe some routing approaches with different algorithms. For the meta-routing layer, we talk about three main modules: *server placement*, *cache organization* and *server selection*.

In Chapter 4, a new QoE-based routing algorithm for the routing layer of a CDN architecture is discussed. The routing algorithm is based on the Q-learning algorithm, an algorithm of reinforcement learning.

In Chapter 5, a new QoE-based server selection method is presented. The method is formalized by the multi-armed bandit (MAB) problem. We also show experimental results of this method.

In Chapter 6, some conclusions and the future challenges of the presented methods are discussed.

1

Network Control Based on Smart Communication Paradigm

The increasing demand for customized and personalized services opened a new overall challenge to find scalable and sustainable solutions for the ever growing smart communications field, which supports different kinds of services, for a wide variety of future next-generation networks applications and cloud computing technologies. To address this challenge, new and cross-disciplinary approaches are required to optimize the distribution of media data which needs to explore how smart communications, built from user perception and feedback, affect protocols, global design, equipment, algorithms, paradigms, power consumption, etc., for a large family of applications using different network technologies. The key observation behind this chapter is that the addressed problem raises many interesting challenges when considered from a control theoretic perspective. First, the general framework is presented. Then, the main inter-disciplinary views are described in detail.

1.1. Motivation

The last two last decades of the 20th Century were driven by the emergence and evolution of the Internet and its use for human communications. There is no doubt that in the 21st Century, the concept of Internet of Things (IoT) and its applications will have a key role in the way we understand our society. The IoT describes the trend for environments, buildings, vehicles, clothing, portable devices and other objects to have a digital representation and the ability to sense, use or exchange information. IoT makes things more interesting by connecting real-world objects, places and people through the digital world. Small objects connected together as part of the IoT today are considered one of the main challenges and the business revolution for the coming years. A widespread use of such connected objects will influence people, societies and businesses. IoT will transform how we live in our cities, how we travel, how we manage our lives sustainably, how we age and how we engage with services and entertainment.

Over the years, the continuous technological evolution and the development of new applications and services have steered networking research toward new problems, which have emerged as the network evolves with new features towards what is usually referred to as the future internet, which has become one of the basic infrastructures that supports the world economy nowadays. In fact, there is a strong need to build a new network scenario, where networked computer devices are proliferating rapidly, supporting new types of services, usages and applications: from wireless sensor networks and new optical network technologies to cloud computing, high-end mobile devices supporting high-definition (HD) media, high-performance computers, peer-to-peer networks and various platforms and applications. This new network scenario is fueling research in the area of new network architectures that consider both the requirements and demands of key emerging applications and services, such as cloud computing and Internet video, and the currently deployed network infrastructures.

Conversely, media distribution over the networks is now growing at a pace that will very soon threaten to change networks to a degree at which the perceived quality will no longer be acceptable. Also, users are switching from operator-controlled IPTV services to uncontrolled over-the-top media applications. High-quality delivery platforms increase the user expectation of quality, equivalent to the quality of traditional broadcast distribution. Furthermore, live TV, for example during sports events, may have a large impact on the amount of traffic in small geographic areas. Various distribution formats, delivery platforms and methods for protecting material leads to many copies of the same material that must be produced and distributed in parallel. Transcoding of content to different formats requires a lot of energy for both calculation and cooling, which is at odds with modern standards of energy efficiency and long-term durability. Also, users become connected to the Internet through different devices (mobile, PCs, HD displays, etc.) and networks (3G/4G/UMTS, ADSL, VDSL, etc.). Devices and networks can vary a lot in their features and capabilities. Hence, each user has a specific context. This situation produces a very heterogeneous environment where an increasing demand for customized and personalized services exists.

Therefore, the overall challenge is to find scalable and sustainable solutions for the ever growing smart communications field, which supports different kinds of services, for a wide variety of future next-generation network applications and cloud computing technologies. To address this challenge, new and cross-disciplinary approaches are required to optimize the distribution of media data. One approach is to introduce and combine methods for more effective storage of the data close to the consumer and various types of peer-supported distribution mechanisms. Decisions on when and where data are temporarily stored should be based both on estimates of expected demand and the final consumer demands on perceived quality. Also, statistical methods for online estimation of consumer demands will be crucial. Many of these issues are complex and can, in the short term, hardly be solved by any single approach. We need to find solutions to deliver network services in the most efficient way to provide users with the best perception while taking into consideration scarce network resources.

We also need to explore how smart communications, built from user perception and feedback, affect protocols, global design, equipment, algorithms, paradigms, power consumption, etc., for a large family of applications (healthcare, subaquatic, vehicular, robotic, economics, etc.) using different network technologies (Sensor Networks, MANET, VANET, Ubiquitous, Virtualization, Data centers, etc.). Indeed, autonomous applications embedded in complex configurations and dynamic environments have experienced a rapid expansion from classical applications where different modular devices, actuators and sensors interact closely. This expansion has considerably impacted the control of a given system in a centralized manner. Current research trends propose new autonomic architecture schemes that manage and control future emerging networks: Sky of clouds, IoT, Smart Grids, Smart Cities, etc. In parallel, the evolution of Internet usage appeals for higher quality guarantees in order to support stringent services. In this context, for example, healthcare and wellness applications such as helping elderly people, assisting dependent people or habitat monitoring in a smart environment, constitute some of the potential scenarios of convergence between autonomous systems and smart network technologies. These applications, which are based on high-level commands, accomplish some specific tasks, reveal new challenges regarding mechanic design, portability, acceptability, power support and efficiency, control theory, etc. In addition to achieving portability and low-power systems, major challenges that substantially limit the efficiency of any autonomous system must be addressed. This new vision will highlight the overlapping of two domains, namely autonomous systems and smart network technologies, devoted to different applications and a large variety of domains.

1.2. General framework

Our vision of a new paradigm is to make interactions first-class objects from the perspective of the user, the application and the network components. This is achieved by analyzing the interaction between the user and the application with quality perception metrics that are used to fix the control/command chain in network components. The idea here is how to integrate these metrics into a control/command chain in order to construct a network system. The key idea is to design an adaptive loop for improving the service quality of a network system by taking into account the end-user feedback. In this model, the perception measurement module evaluates the user feedback and sends the result to the control module. The latter analyzes the received results, makes a decision and sends it to the command module that initiates the chosen action onto the network system. This loop model refers to an adaptive system based on the end-to-end user's perception and results in smart communications between equipment.

The key observation behind this vision is that the addressed problem raises many interesting challenges when considered from a control theoretic perspective. The characteristics of multimedia contents and the network resources vary over time. There are propagation delays in networks and the quality of offered services is still difficult to evaluate. For example, multimedia contents are strongly linked to the observed rate distortion. In fact, techniques borrowed from control theory, ranging from stability and controllability techniques, can be applied to adaptive, predictive, or stochastic control in both centralized and distributed multimedia delivery architectures. Unexplored challenges in control techniques for efficient multimedia delivery will be emphasized.

The use of artificial intelligence tools together with biologically inspired techniques, to control network behavior in real time so as to provide users with the Quality-of-Service (QoS) that they request, and to improve network robustness and resilience, is necessary. The key idea is based on control theory and machine learning techniques in order to support smart communications driven from user's perception to avoid any complete interruption in the whole chain of service treatment.

This is not the first attempt to get rid of the notion of technical control mechanisms in network components based only on network performance. The operators always try to deliver network services in the most cost-effective and resource-efficient manner with assured user perception. Therefore, service providers are switching their focus from traditional control mechanisms to user satisfaction, which is the overall success of a network from the user's perspective. According to Daniel R. Scoggin, "The only way to know how customers see your business is to look at it through their eyes". In fact, the convergence of network technologies has been driven by the converging needs of end users. The perceived end-to-end quality becomes one of the main goals required by users that must be guaranteed by the network operators and the Internet service providers, through manufacturers' equipment. This is referred to as the Quality of Experience (QoE) notion, which is commonly used to represent user perception. QoE is not a technical metric, but rather a concept consisting of all elements of a user's perception of the network services such as those found in the field of affective computing. To ensure good end-user perception, service providers have encountered QoE challenging issues, such as QoE monitoring, measurement, diagnosis and management. Therefore, all of network functions such as admission control, access network selection, routing, resource allocation, transmission control and server selection are expected to be adaptive to user QoE.

This vision differs from others by focusing on control theory and machine learning techniques in order to support smart communications to avoid any complete interruption in the whole chain of service treatment based on the user's perception. The bio-inspired model is one of the possible ways in which machine learning addresses the large-scale and distributed issues in new environments such as massively large-scale, wide-area computer networks and mobile *ad hoc* networks. These new environments also present new challenges because they are highly dynamic, sometimes unreliable, and often span over a wide area. The aim here is to demonstrate that these models based on living organisms can effectively organize themselves, in large numbers of unreliable and dynamically changing components, into structures that implement a wide range of functions. In addition, most biological structures exhibit several favorable properties such as robustness to failures of individual components, adaptability to changing conditions and dynamic scenarios and their low reliance on explicit central coordination.

Many works have been done based on continuous learning paradigm. For example, in our early work [MEL 05], we demonstrate that the approaches based on central and explicit control over the system are also not suitable for large-scale distributed networks because they are not scalable. In addition, centralized approaches introduce a single-point-of-failure, which should be avoided whenever possible. The article introduces an adaptive model that is very useful in tracking a phenomenon that evolves over time. The problem was naturally formulated as a dynamic programming problem, which, however, is too complex to be solved exactly.

The most important challenge is the use of the concept of a state-dependent model based on a global optimization cost function whose parameters are learned from the real-time state of the network with no *a priori* model and applied in several domain applications. This kind of model is based on the fact that it is capable sampling, estimating and building the network environment's model of pertinent and important aspects, such as type of traffic, QoS policies and resources. It is based on the trial/error paradigm combined with swarm adaptive approaches. In a more recent work, we developed a global system that uses a model which combines both a stochastic planned pre-navigation for the exploration phase with a deterministic approach for the backward phase.

We have studied and introduced the concept of control mechanisms driven by user perception. In fact, we observed that the most important challenge of developing an efficient QoS mechanism is to resolve the NP-completeness complexity problem in the presence of more than two non-correlated criteria [GAR 79]. The solution proposed is to integrate the user's perception to control the time-evolving optimization cost function in order to maintain a certain satisfaction degree of end users, which is the objective of all network systems, without optimizing all network technical parameters.

The development of individual perception estimator agents, including the human perception models, the service models and the subjective test methodology, and the absence of applications address the problems of smart communications outlined in section 1.1: scalability can be achieved by designing metrics that support large data sets; platform diversity is supported by the ability to use services that take advantage of the platform's interaction resources and by a natively distributed architecture; flexibility is supported by the independence between metrics and the services that they operate on; and heterogeneity is supported by interconnecting different kinds of quality estimators.

1.3. Main innovations

The following sections describe several main inter-disciplinary views in order to accomplish the main tasks of the development of this new vision: developing at the theoretical level the appropriate unified formal model based on bio-inspired modeling and knowledge distribution in order to construct a scalable and robust environment with low complexity for large-scale dynamic networks; studying at the empirical level why and how user perception can be quantified, analyzed and influenced; creating at the conceptual level a general framework for protocol stacks dedicated to smart communications featuring multiple technologies, devices and users; and building at the engineering level the appropriate model of programming abstractions and software architecture to develop a full-scale framework.

1.3.1. *User perception metrics and affective computing*

As a critical measure of the end-to-end performances at the service level from the user's perspective, the user's perception is a kind of metric for the system design and engineering processes that can only be measured dynamically at the very end of any transmission activity.

To clarify the above definition, many standards organizations have studied this concept. We cite here three relevant standardization organizations:

– ITU-T: Study Group 12 is the main group studying the QoS and QoE in the ITU Telecommunication Sector (ITU-T). Some important recommendations include: ITU-T G.1010 [REC 01a], ITU-T G.1030 [REC 05, REC 11b] and ITU-T G.1070 [REC 07].

– Broadband-Forum [BRO 95]: the user's perception is considered as the measure of overall performance from the user's perspective.

– TeleManagement-Forum (TM Forum) [TMF 88]: TM Forum considers a set of key quality indicators (KQIs) and key performance indicators (KPIs) as measurements of perceived quality rather than network performance.

In fact, even though user perception is difficult to predict due to its subjective nature, defining a set of metrics is necessary for the assessment of overall service quality. One way to achieve this is using the top–down approach in which the factors that influence user perception are determined at the beginning. Then, this information is used to generate operating requirements. Finally, a method to constantly evaluate these factors and improve them is created based on reference measurements (absolute and/or relative evaluations), preservice and in-service tests (achieved offline or online), and problem detection (image quality, audio quality, video quality, interruption, task completion, time, etc.). Also, the perception control procedure, i.e. mapping of the user's perception onto the network performance, is an important challenge. To achieve this goal, we have to identify the relationship between network performance metrics and their effects on user perception [ARO 12].

Also, a critical step in managing and evolving the network and business is to identify and measure the right KQIs. When identifying KQIs, there will be as many different expectations as there are users, but most of these expectations can be grouped under main categories to be defined. For example, in network services, the reliability dimension has some KQIs as follows: service availability (anywhere), service accessibility (anytime), service access time (service setup time) and continuity of service connection (service retainability). On the other hand, for telecom applications, KQIs could be as follows: connection establishment success rate, connection drop rate, handover success rate, throughput and block error rate. In fact, the KQIs are measured for each service in order to calculate an estimate of each service score.

The other crucial question here is to develop systems and devices that can recognize, interpret, process and simulate human affects. It is an interdisciplinary field spanning computer sciences, psychology and cognitive science [PIC 97]. Detecting emotional information begins with passive sensors that capture data about the user's physical state or behavior without interpreting the input. The data gathered are analogous to the cues humans use to perceive emotions in others. For example, a video camera might capture facial expressions, body posture and gestures, while a microphone might capture speech. Other sensors detect emotional cues by directly measuring physiological data, such as skin temperature and galvanic resistance. Recognizing emotional information requires the extraction of meaningful patterns from the gathered data. This is done using machine learning techniques that process different moralities using speech recognition, natural language processing or facial expression detection, and produce either labels (i.e. "confused") or coordinates in a valence-arousal space.

Conversely, different sources of uncertainty and imprecision may arise with these collected measurements. Such imperfections may be due to the imprecision of many environmental aspects: signal, data link, network, etc. For example, it may be due to communication links that are unreliable, either due to operational tolerance levels or environmental factors. The theory of belief functions can be used here to represent, correct and reason with partial information [SHA 13].

1.3.2. *Knowledge dissemination*

As explained earlier, the objective here is to transform the current network centric view to a future user-centered view. To make appropriate decisions, such a network needs to collect a quantity of knowledge, which forms a new knowledge plane. The knowledge plane must be created autonomously, continuously and dynamically. The knowledge must be acquired in both a vertical manner (knowledge emanating from different layers of the same node) and in a horizontal manner (knowledge coming from multiple nodes). However, the most important factor is to have an entire distributed knowledge plane [LI 08]. To do this, knowledge must be stored (physically and logically) in different parts of the network.

This knowledge plane must meet three main features [SOU 12]:

– The knowledge must be collected in real time using various monitoring tools on different network layers.

– An analysis and learning component can make the best decisions based on the context.

– The knowledge must be shared among different nodes.

It is clear that the knowledge dissemination in the network remains an open problem. In fact, previously realized works on the autonomic paradigm are mostly based on a multi-agent. An agent is a computer system located in an environment and acting autonomously to achieve the objectives for which it was designed. By distributing agents on each equipment, the system can handle local problems locally, based on all acquired and taught knowledge [KEP 03]. In most situations, local problems are more easily resolvable than global problems, especially as a local problem is detected much faster than in an approach called centralized. Obviously, this type of architecture is not realistic because the knowledge dissemination will generate a very high overhead.

From a different point of view, many service quality parameters influence end-user satisfaction. Hence, designing Service-level agreement (SLA) oriented resource management schemes are required for service operators that provide attention to each customer. For instance, this can be achieved by enabling communication to keep end users informed and obtain feedback from them, increasing access and approachability for end users and understanding the specific needs of each end-user flow.

Moreover, service requirements of users can change over time and thus may require updating of original service requests. For example, the resource management algorithms will be able to self-manage the reservation process continuously by monitoring current service requests, predicting future service requests, and adjusting schedules and prices for the current communication.

Hence, self-configuring components to satisfy new service requirements are needed, so that more autonomic and intelligent providers can better manage the limited supply of resources with dynamically changing service demand. For end users, the system acts on their behalf to select suitable providers and negotiate with them to achieve ideal service contracts. Thus, providers also require autonomic resource management to selectively choose appropriate requests to accept and execute depending on various operating factors, such as the expected availability and demand of services (both current and future), and existing service obligations.

1.3.3. *Bio-inspired approaches and control theory*

Many studies have been conducted over the last decade for an alternative paradigm that would address the integration of dynamic criteria evolving in time with uncertainty. The most popular formulation of the optimal distributed optimization problem in a network is based on a multiparameter flow optimization whereby a separate objective function is minimized with respect to the types of flows subjected to multiparameter flow constraints. However, due to their complexity, high processing overheads, a few proposed schemes could be accepted in real life. For a network node to be able to make an optimal control decision, according to relevant performance criteria, it requires not only up-to-date and complete knowledge of the state of the entire network but also an accurate prediction of the network dynamics during propagation of the message through the network. This, however, is impossible unless the control algorithm is capable of adapting to network state changes in almost real time. So, it is necessary to develop a new intelligent and adaptive control algorithm. This problem is naturally formulated as a dynamic programming problem, which, however, is too complex to be solved exactly. In these adaptive paradigms, the environment is modeled as stochastic (especially links, link costs, traffic, congestion, etc.), hence control algorithms can take into account the dynamics of the network. However, no model of dynamics is assumed to be given. This means that these approaches have to sample, estimate and perhaps build models of pertinent aspects of the environment in order to support a framework that addresses the identified trade-offs needed for an overall control mechanism.

Many different approaches based on bio-inspired paradigms, such as cognitive packet networks [GEL 04], ant colony optimization and reinforcement learning paradigm [SUT 98, MEL 11a], exist in the literature, and the objective is to see how to reduce their complexity. In addition, such a network system must also design and implement intelligent and adaptive optimizing control algorithms to support irregular and differentiated flows, and take into account the network state and its dynamics. The use of these bio-inspired models is one of the possible ways to address the large-scale and distributed issues [DRE 10]. It is well known that living organisms can effectively organize themselves in large numbers of unreliable and dynamically changing components into structures that implement a wide range of functions. In addition, most biological structures exhibit several favorable properties such as robustness to failures of individual components, adaptability to changing conditions and dynamic scenarios, and their low reliance on explicit central coordination. Currently, the main research direction in the bio-inspired modeling area focuses on the exploration of solutions that enable the construction of a scalable and robust environment with low complexity for large-scale dynamic networks. Such an objective can be met not only by relying on the decentralization of functions or their self-organization but also by depending on the methods, techniques and tools for data retrieval and analysis and learning, all of which participate in the global decision process (thereby simplifying and minimizing human actions and interventions).

1.4. Conclusion

The addressed problem raises many interesting challenges when considered from a control theoretic perspective. The characteristics of real-time service contents are time-varying as are the network resources. Control theory offers interesting solutions for several problems related to the control of networks, such as the stability of protocols in order to guarantee a level of performance on all the space of uncertainties. This new vision will focus on techniques borrowed from control theory, ranging from stability and controllability techniques, to adaptive, predictive or stochastic control in both centralized and distributed network delivery architectures.

2

Quality of Experience

First, this chapter presents the definitions and explanation of importance of Quality of Experience (QoE) in the network context. The Quality-of-Service (QoS)/QoE relationship and some QoE measurement methods will be also discussed. The state of art of different approaches for each layer will be described in detail.

2.1. Motivation

Nowadays, next-generation networks (NGN) trend includes deploying the fixed mobile Internet Convergence (FMIC) and migrating the traditional telecom networks to Internet Protocol (IP) technology. NGN will employ multiple networking technologies for the best network service. Network operators and service providers try to simplify network operations and management in order to offer multiple services over a single network. While NGN network experts are going to employ a common network layer protocol in core networks to accomplish the current network services, the access networks will use numerous technologies, such as WLAN, WPAN, Ethernet cable, DSL, 2G/3G/4G, LTE, WiMAX, UWB and optical fiber, to meet the diversified requirements of end users [ZHA 11]. Using a network environment with multiple operators and multiple networks, end users expect to use a heterogeneous wired and wireless high-bandwidth ubiquitous network access and diversified services. Hence, the end users have a multitude of offers that decrease service prices. Consequently, the competition between network providers increases. So, the customer is today in a strong position, being able to select between different competing providers. Apart from similar pricing schemes, which are a useful decision aid for users, their choices are also influenced by the expected and experienced quality. So, the interest of service providers in how users perceive usability, reliability, quality and price-worthiness has increased. As a result, the QoE notion has been introduced to help service providers to take into account user perception and satisfaction.

As mentioned in the Introduction, this book presents an adaptive end-to-end (E2E) QoE control model and applies it to a real infrastructure, the content distribution network (CDN), to construct a QoE-based content distribution network architecture (QCDNA). Before presenting our QCDNA, we need to clarify what QoE is and how important it is in the network context. Afterwards, it is also necessary to describe the CDN architecture to show the need of applying an E2E QoE model. Therefore, in this chapter, we survey the QoE notion in the networking context.

2.2. QoE concept

The era of user-centered network service has begun, where user perception plays a key role in the entire system. The quality perceived by end users when using a service is called QoE. Taking image processing as an example, the question of which features are included in a multimedia service becomes how well such features are addressed and the impact they have on end users. The data usually pass through many processing phases before delivering to end users host. These phases include: acquisition, processing, coding, transmission, enhancement, decoding, etc. The quality of multimedia data may be affected by each of these phases. This affection decreases the E2E QoE. There are also some other factors that impact on the QoE such as usability, human factors and context.

Traditionally, service providers carry only the notion of QoS, which objectively measures and guarantees characteristics of service from the provider perspective. QoS metrics take into account system components' characteristics, not the human perception. Over the last few years, QoE has brought a new look to E2E quality in multimedia systems. QoE represents aspects related not only to subjective perception, but also user behavior and needs, appropriateness, context, usability and human factors of the delivered content.

The first concern refers to what the QoE is. There are different definitions of QoE. According to International Telecommunication Union (ITU) [ITU 65], QoE is defined as "the overall acceptability of an application or service, as perceived subjectively by the end user". In fact, QoE is more a qualitative measure than a quantitative measure. In PC Magazine [PCM 94], QoE is defined as "a subjective measure of performance in a system. QoE relies on human opinion and differs from quality of service (QoS), which can be precisely measured". According to Fiedler *et al.* [FIE 10a], if a user does not feel that (s)he is getting "value for money", i.e., good-enough QoE for what (s)he is paying for, (s)he might be ready to abandon the particular service provider and/or network operator in question.

To clarify the above definitions, many standards organizations have studied the concept of QoE. We cite here three relevant standards organizations:

1) *ITU-T*: Study Group 12 is the main group studying the QoS and QoE in the ITU Telecommunication Sector (ITU-T). Some of the important recommendations are as follows:

– *ITU-T G.1010* [REC 01a] constructs a model for multimedia QoS categories from an end user viewpoint. This recommendation studies not only the key parameters impacting the user such as the delay, delay variation and information loss, but also the performance targets for different applications that would meet user expectations. The intent of this recommendation is to provide guidance on the key factors that influence QoS from the perspective of the end user.

– *ITU-T G.1030* [REC 05] proposes a framework for obtaining the IP network performance, evaluates the application performance and assesses user satisfaction with the E2E performance. The recommendation shows a model of Web browsing. This guideline has three steps: network performance evaluation, application performance evaluation and perceptual performance evaluation. The third step introduced the user perception. This model is similar to the E-model [REC 11b] that makes a link from end user experience of network layer to application layer.

– *ITU-T G.1070* [REC 07] provides a model that assesses videophone quality of the QoE/QoS planners. This model is used to guarantee the user satisfaction with E2E service quality. It has three functions: *video quality assessment*, *speech quality assessment* and *multimedia quality integration functions*. The model is based on specific terminals and environments.

2) *Broadband-Forum* [BRO 95]: besides their focus on triple-play applications (i.e. video, audio and best-effort data), Broadband Forum also takes into account the QoE/QoS relationship. With a similar idea on QoE definition of the ITU-T, they consider QoE as the measure of overall performance from the user perspective. They try to define a clear relationship between these two concepts so that we can evaluate QoE from QoS measurements. Furthermore, they propose the idea of determining the required network parameters with a target of QoE. The Broadband Forum provides a complete methodology and some specific metrics (e.g. delay, bandwidth and loss rate), but they do not show how QoE would be affected if such requirements were not met.

3) *TeleManagement-Forum* [TMF 88]: TM Forum uses service level agreement (SLA) to assess QoE. TM Forum considers key quality indicators (KQIs) and key performance indicators (KPIs) as measurements of perceived quality rather than network performance. They also focus on QoE management and try to find an exact definition of E2E QoE in an entire network system.

All these standards organizations focus on providing a clear and uniform definition of QoE and QoS. Moreover, they also try to establish the relationship between QoS and QoE in order to contribute to the QoE measurement and management.

Classification is necessary for understanding any concepts. QoE is not an exception. We classify this notion into three different classes (Table 2.1): *QoE as QoS extension*, *QoE for management systems* and *QoE as a hedonistic concept*. We have chosen this method of classification because these three approaches can cover the whole QoE notion. Indeed, what makes us interested in considering the user perception concept are the following:

– QoE as QoS extension: this refers to the fact that QoE is generally viewed as an extension of a known concept, QoS. From a historical point of view, the QoS metric first came to designate a set of techniques to ensure the routing of network traffic such as voice or sensitive multimedia applications. Since then, QoS has been highlighting the performance improvement of network systems. The QoE concept appeared more recently. As presented, a comprehensive approach to quality (measured E2E), QoE directly affects end users and is considered as an extension of QoS.

– QoE for management systems: a management system of network services is responsible for ensuring availability, performance and efficiency of all modules in the framework. Over the last few decades, the network management systems have only been based on technical parameters such as delay time, loss-rate and bandwidth. However, satisfying all these metrics does not ensure the success of the system. Service providers unintentionally forgot the most important factor: the user experience (UE), which plays a critical role in the whole system. That is why, we consider QoE as the key factor for network management system.

– QoE as a hedonistic concept: in the context of QoE, the hedonistic concept is related to the pleasure and satisfaction of the end user when using a service. QoE represents the overall end user satisfaction with a service. Even though QoE is not always numerically quantifiable, it is the single most important factor for assessing the UE. Hence, QoE can be viewed as a hedonistic concept.

2.3. Importance of QoE

E2E-QoE measures the degree of satisfaction of services perceived by the end user both objectively and subjectively. It has become a major issue for telecommunication companies because their businesses are highly dependent on customer satisfaction, and thus, the average revenue per user can only be increased by value-added services, which requires taking into account the complete E2E system effects. E2E-QoE involves the entire E2E path chain including the Quality of Design (QoD)[1] of terminal equipment and QoS supported by network components. In the context of communication services, E2E-QoE is influenced by content, network, device, application, user expectations and context of use. Other external

1 The QoD is the quality perceived by end user when using end-devices. In other words, QoD is how end user is satisfied with design and usability of end-devices.

factors that can also have an impact on E2E-QoE include user's terminal hardware, mobility, and the importance of the application.

QoE approaches	Descriptions
Extension of QoS	QoE is considered as an extension of QoS. That reflects the QoS influences on end users' perception.
Key factor for management systems	QoE is considered as a key factor for management systems. It reflects the user's perception, and has vital importance for management tasks.
Hedonistic concept	QoE determines the user satisfaction when using a service. The key idea of the hedonistic concept is that pleasure is the only factor that has intrinsic value. This concept, in the context of QoE, is used to evaluate service in terms of how much pleasure and how little pain end users experience.

Table 2.1. *Three QoE approaches*

Technical metrics such as packet delay, loss rate and jitter can represent QoS notions, and for some of the high-quality conferencing applications, these QoS metrics certainly become important. However, the QoS parameters are not the only crucial factor that determines service quality. Therefore, QoE aims to estimate or measure what happens in the customer's mind, i.e. service quality perceived by the customers. As we know, simply improving technical metrics by itself would never be enough to prevent customers from giving up the service. Instead, operators should use some metrics that can react in advance and improve customer perception, i.e. QoE. As a result, network providers who provide good QoE have a significant competitive advantage, while providers who ignore the importance of QoE may suffer unnecessary costs. [SOL 06] confirm this point in showing statistical results: approximately 82% of customer defections (churning) are due to frustration over the product or service and the incompetence of the provider and 90% of users will not complain before defecting. The results indicate the necessity of a strategy to manage and improve QoE proactively for customer satisfaction.

Today, with the highly competitive environment of various service providers, end users have the option to choose the best service provider. As a consequence, the service availability is not enough. In this context, the service providers must improve those services so that users can enjoy a rich experience with a reasonable price. In addition, data traffic is today no longer just the best-effort traffic. To meet user needs and requirements, many applications of data traffic need high QoS. Thus, many standards organizations (e.g. ITU, Broadband Forum, TeleManagement-Forum as mentioned in the previous section) have proposed many classifications of services and QoS parameters. However, they do not clearly capture the real QoE. The most important difference between QoS and QoE is that while QoS just focuses on measuring performance from network perspective, the QoE focuses on how well a service meets the user's expectations. Therefore, the challenge for service providers is to take into account both QoE and QoS. In many cases, acceptable QoS does not always infer acceptable QoE. For example, in a session of VoIP [BAT 10], the packet overflow is monitored at the aggregate queue level. Assume that an edge router has an inadequate buffer size. As the VoIP traffic volume is low compared to other types of traffic, the aggregate buffer statistics give an acceptable QoS. Despite this, the QoE in this case is poor because of dropped packets. So, the VoIP subscribers are unsatisfied.

2.4. QoE metrics

Even though QoE is difficult to predict due to its subjective nature, measurement of QoE is necessary for the assessment of overall service quality. Therefore, we need an effective strategy to measure QoE as realistically as possible. One way to achieve this is by the top–down approach. In the top–down approach, the factors that influence user perception are determined at the beginning. Then, this information is used to generate operating requirements. Finally, a method to constantly evaluate these factors and improve them is created.

We can group the expectations obtained in the initial phase into two main categories: *reliability* and *quality* [SOL 06]. Quality in this case refers to the QoS and application software features. KPIs for the *reliability dimension* in the network service include:

– service availability (anywhere);

– service accessibility (anytime);

– service access time (service setup time);

– continuity of service connection (service retainability).

KPIs for the *quality dimension* include:

– quality of session;

– bit rate;

– bit rate variation;

– active session throughput;

– system responsiveness;

– E2E delay;

– delay variation.

The KPIs are measured for each service in order to calculate an estimate of each service score. Based on the comparison between these scores and the initial "QoE targets" which were set before measuring each KPI, the values of each of these metrics will be mapped to assess the level of impact on the actual QoE.

Some KPIs will be totally irrelevant in one case while being the most important in another. It all depends on the type of service application the user is running. For example, important characteristics of voice and video conferencing are time relation preservation, guaranteed resource allocation without retransmissions and real-time traffic. In applications such as Web browsing, remote server access and interactive gaming, however, characteristics such as request/response pattern of use, transmission delay, dynamic resource allocation and best-effort traffic should be considered more.

Another way to classify QoE metrics is based on *subjective* or *objective* methods. Subjective methods are conducted to obtain information on the quality of multimedia services using opinion scores, while objective methods are used to estimate the network performance using models that approximate the results of subjective quality evaluation. Subjective parameters assess how audio and video data are collected from users, namely what their opinion is about the quality of sequences of audio/video. A typical example of qualitative metrics (subjective) is the mean opinion score (MOS) [REC 06], which was first standardized by the ITU. With this metric, the system quality is assessed subjectively from the mean opinion of a panel or group of users. Each user rates the service on a five-point scale: 1 = bad, 2 = poor, 3 = fair, 4 = good and 5 = excellent. The MOS is used typically in subjective tests of audio listening, where a number of people asses the quality of the audio. Another example of a qualitative metric is the R-factor [REC 11b], which can be used in a manner similar to MOS. Subjective evaluation of speech quality uses the R-factor in voice transmission models. Examples of subjective tests were summarized in ITU-R Rec. BT.500-11 [REC 02] and ITU-T Rec.P.910 [REC 99] and cited in [CHI 11].

– Double Stimulus Continuous Quality Scale (DSCQS) [REC 02]: in this test, the evaluators watch the reference and processed video twice. They then rate the video quality on a continuous quality scale of 0–100 (with 0 being *bad* and 100 *excellent*).

– Single Stimulus Continuous Quality Evaluation (SSCQE) [REC 02]: the evaluators watch only the processed video sequence with long duration (approximately 20–30 min). They use a slider to rate the instantaneous perceived quality on the DSCQS scale of *bad* to *excellent.*

– Absolute Category Rating (ACR) [REC 99]: it is similar to SSCQE with only the processed video being shown to the evaluators. However, it is very common to include hidden references that are rated in the same way as those which are processed [ROH 00]. The evaluators rate the overall video quality using a discrete five-level scale ranging from *bad* to *excellent.*

– Pair Comparison (PC) [REC 99]: in this method, the evaluators watch test clips from the same scene but under varying conditions. These clips are paired in all possible combinations and screened to the evaluators for preference judgment about each pair.

– Double Stimulus Impairment Scale (DSIS) [REC 02]: unlike the DSCQS, in the DSIS, the evaluators only have to watch one presentation sequence. The evaluators rate the sequences on a discrete five-level scale ranging from *very annoying* to *imperceptible*.

However, subjective QoE measurement is time consuming, tedious, expensive and is not particularly applicable in a production environment. Instead of directly collecting quality information, objective methods can be used to estimate the MOS score. This way, the scores reflecting real-time user perception can be estimated and, immediately, the service providers can utilize the information to maintain the network condition thereby providing best QoE.

The objective quality measurement methods have been classified into the following five main categories [CHI 11, TAK 08]:

– Media-layer models: the QoE is calculated using the speech and video signal. This model does not require any information about the system under testing (e.g. codec type, packet-loss rate). It can be applied to the evaluation of unknown systems.

– Parametric packet-layer models: these models predict the QoE from the packet-header information and do not analyze media signals.

– Parametric planning models: these models use quality planning parameters for networks and terminals to predict the QoE.

– Bit-stream-layer models: in these models, encoded bit-stream information and packet-layer information are used to measure QoE.

– Hybrid models: these models are a combination of previously mentioned technologies.

Among these categories, the media-layer objective quality measurement methods are subdivided into three types: (Figure 2.1) full reference (FR), reduced reference (RR) and no reference (NR). Taking video signals as an example, these subtypes can be explained as follows:

– FR: the degraded signal is compared pixel by pixel with the original signal. It is a very detailed comparison that requires the original signal. Sending the reference signal requires considerable computing resources. In addition, this method may not be applicable because the reference image is often not available. Some of these metrics have been standardized by the ITU [REC 04a, REC 04b, REC 08b].

– NR: this is a video analysis (stream) on receipt without comparing it to the original signal. Quality information is extracted from the original signal, as no reference is available. These methods are very small, making them very suitable for analysis in real time. However, KPIs output is low, especially compared to “Full Reference” methods. In addition, NR metrics cannot distinguish the difference between quality problems of the very reference and additional disturbance by the network. This method is an extremely difficult task (especially general-purpose metrics that are applicable to a wide variety of image distortion types).

– RR: these methods actually build upon representative parameters (typically statistical values) that allow for the quantification of the change of quality between the original and the distorted version. This method provides a solution that lies between FR and NR models. With this method, the perceptual quality of distorted images is predicted based on partial information of reference images. Some of these metrics have been standardized by the ITU [REC 08a, REC 10, REC 11c].

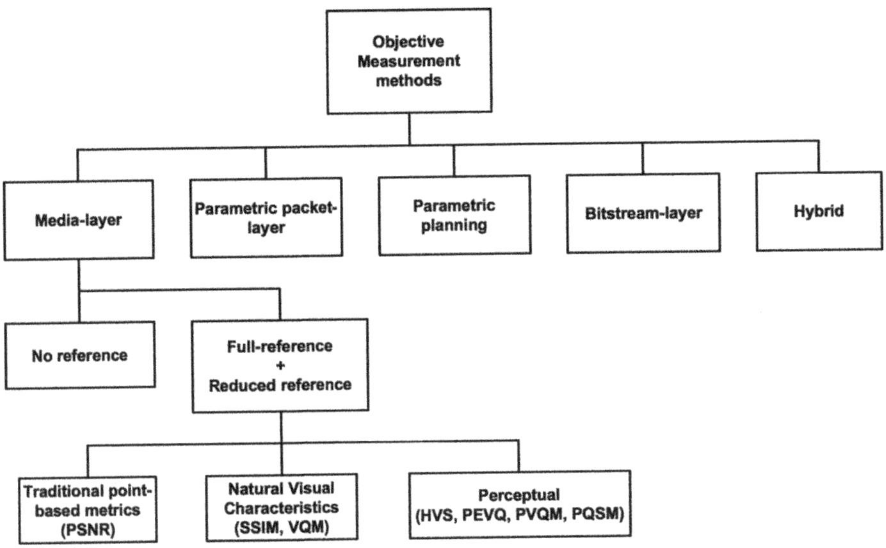

Figure 2.1. *Classification of objective measurement methods*

Figure 2.1 shows us the classification of objective quality measurement methods. The FR and RR are classified into traditional point-based metrics (e.g. peak signal-to-noise ratio (PSNR), mean square error (MSE)), natural visual characteristics oriented metrics (e.g. structural similarity (SSIM), video quality metric (VQM)) and perceptual oriented metrics (e.g. human visual system (HVS), perceptual evaluation of video quality (PEVQ), perceptual evaluation of speech quality (PESQ), perceptual quality significance Map (PQSM)). We now present in detail some representative objective metrics:

– PSNR [HUY 08] is the proportion between the maximum signal power and the corruption noise power. The original data are called the signal, and the error is noise in this case. However, this metric has several limitations and it can only be used as a measure of quality of reconstruction of loss compression codecs (e.g. for image compression).

– SSIM [WAN 04] is a metric used for measuring the similarity between two images. It is an FR metric. The image quality measurement is based on a distortion-free image as a reference.

– PEVQ [REC 08b] takes an FR approach to provide MOS scores of the video quality for IPTV, streaming video, mobile TV and video telephony. This method analyzes the degraded video signal through the network for evaluating the degradation. PEVQ has become part of ITU-T Recommendation J.247.

– PESQ [RIX 02] is a method for evaluating speech quality autonomously as the experience of a telephony system user. It is standardized as ITU-T recommendation P.862 [REC 01b].

– Pseudo-subjective quality assessment (PSQA) [RUB 05] is a measurement method that uses a learning tool to know how users perceive when receiving data. The authors use a random neural network model, which is a network that has positive and negative customers.

– IETF proposed the media delivery index (MDI) [WEL 06], a metric that is based on parameters such as a measure of deviation from nominal flow rates, traffic jitter, and packet loss measurements for a particular flow, and does not consider the codec or video information in its quality assessment process.

The QoE measurement possibility helps operators to maintain networks' performance in the overall level of customer satisfaction in terms of reliability, availability, scalability, speed, accuracy and efficiency.

2.5. QoE measurement methods

We now survey several QoE measurement models that have recently been proposed in the literature. Goong *et al.* [GOO 09] proposed a pentagram model for measuring QoE with five factors: integrality, retainability, availability, usability, and instantaneousness. The authors defined a model of experience taking into account these factors in quality evaluations. The model was applied to measure experience of a VoIP service. QoE KPIs and the most important measures were defined as the following:

– service integration (delay, jitter and packet loss ratio) (symbol: a);

– service retainability (service interruption ratio) (symbol: b);

– service availability (the success ratio of user access service) (symbol: c);

– service usability (service usability) (symbol: d);

– service instantaneousness (the response time(s) of establish and access service) (symbol: e).

Taking each factor into account, the authors propose a formula (equation [2.1]) to calculate QoE:

$$QoE = \frac{1}{2}\sin(\lambda)(ab + bc + cd + de + ea) \quad [2.1]$$

where λ represents the 72-degree angle between the two sides in the QoE pentagram model. This model facilitated QoE measurement during a service use process in order to improve and represent new methods to evaluate the perception of VoIP services.

Kim *et al.* [KIM 08] propose a method for the objective QoE measurement using traditional QoS parameters. The QoS and QoE correlation model was described, and the QoE evaluation method using QoS parameters was studied in the converged network environment. The QoS parameters associated with QoE evaluation and the service classes include delay (D), jitter (J), loss rate (L), error rate (E), bandwidth (B) and signal success rate (S). The QoS–QoE correlation model was expressed as (equation [2.2]):

$$QoS = F(D, J, L, E, B, S) \quad [2.2]$$

$$QoE(QoS) = K\left\{\frac{(e^{QoS-\alpha}) + e^{-QoS+\alpha})}{(e^{QoS-\alpha}) + e^{-QoS+\alpha} + \beta)} + 1\right\} \quad [2.3]$$

where, α is the QoS quality class of the network level. The QoE class measured by the QoS quality parameter of a network level is mapped similarly to the existing MOS grade with a five-point scale. β is determined according to the class of service as the grade of service constant. K is the scale constant to show the satisfaction with the used service. Using the QoS information measured on a network level, authors explain the proposed QoS–QoE correlation model for the objective QoE.

Another approach by Moebs [MOE 08] takes into account the trade-offs between QoS parameters and the flow requirements for e-learning, and aims to develop better evaluation methods for QoE using QoS–QoE correlation. The authors proposed an adaptive multimedia e-learning system that consists of a user model, a domain model and an adaptation model. The domain model represents the concept of the subject domain and usually, these concept structures are described as concept maps, semantic networks or concept graphs. The user model represents general characteristics of the user such as location, preferences for devices, previous knowledge, knowledge state and learning goals. The adaptation model uses adaptation rules to connect the two previously outlined models. Flow-related and learning-related aspects are the principal elements of the QoE, and are both influenced by QoS. This relationship is represented by the following function:

$$QoE = f(QoL(QoS), QoF(QoS)) \quad [2.4]$$

QoS is based on packet loss, delay and jitter in background applications, video, audio and data. Quality of Learning (QoL) represents the aspects of the user model (feedback, learning styles, assessment of the previous knowledge) and the domain model. Quality of Flow (QoF) is based on feedback, interaction, the balance of skill and challenges, which contains use of technology and emotions.

Hamam *et al.*[HAM 08] propose a taxonomy that is modeled by a mathematical model where the QoE is calculated as the weighed linear combination of the QoS and UE for a particular haptic user interface:

$$QoE = \zeta \times QoS + (1 - \zeta) \times UE \quad [2.5]$$

where ζ controls the relative weight given to the QoS parameters compared to the UE parameters. UE is a linear combination of perception measures, rendering quality, physiology measures and QoS measures. If the quality factors are restricted between 0 and 1, then the overall QoE will also have a value between 0 and 1. To achieve this condition, the sum of constant coefficients must be equal to 1.

The four previous approaches have shown different methods for measuring the QoE. However, all these methods are based just on the linear combination of QoS metrics (objective elements) and on the QoS/QoE correlation, from which the authors calculate QoE. It seems that they have ignored subjective perception, an indispensable factor of the QoE concept. QoE measurement methods must reflect the user's perception and satisfaction. Certainly, the QoE of audio or video media should be evaluated in subjective terms. Subjective quality measurement is the most basic method of evaluating subjective quality. This deficiency motivates us to use a hybrid tool that mixes the two types of measurement: the subjective and objective method.

The meaning of QoE in a network context is so important that we have to find out the relationship between the QoS and the QoE. Actually, understanding the relationship between a familiar concept (QoS) and the new concept (QoE) is the best way to understand this new concept. The next section clarifies this.

2.6. QoS/QoE relationship

The notion of QoS [REC 11a, REC 01a] has been discussed for more than a decade. It concerns essentially the technical view on service quality. The QoS architectures (e.g. integrated services or differentiated services) are used to pave the way for high-quality real-time services such as VoIP or video streaming. However, the lack of UE consideration raises unresolved issues. As a result, in order to redirect the focus toward the end user and quantify the subjective experience gained from using a service, the notion of QoE has been introduced to describe quality as perceived by the human user.

QoS is defined as "ability of the network to provide a service at an assured service level" [SOL 06], or "capability of a network to provide better service to selected network traffic described by the following parameters: delay and jitter, loss probability, reliability, throughput and delivery time" [MAR 07a]. These definitions show that QoS is fundamentally a technical concept, which usually has little meaning to users. There are also some QoS definitions that correlate with the QoE: "collective effect of service performance which determines the degree of satisfaction of a user of the service" [REC 94], "a set of qualities related to the collective behavior of one or more objects" [ISO 98], or "the network's capability to meet the requirements of users and applications" [KIL 02].

Compared to QoS, the notion QoE means the user perception about the quality of a particular service or network. It is expressed by human feelings like "good", "excellent" and "bad".

In order to clarify the relationship between QoS and QoE, we observe a quality chain of E2E services (Figure 2.2) where QoE is influenced by all intermediate aspects in the E2E chain, such as Quality of Perception (QoP), QoD and QoS. QoP is an aspect related to user's perception and to his personal profile (e.g. age and sex) and his psychological status (e.g. happy, nervous and uninterested). QoD, which depends on designers and manufacturers, is defined as the quality perceived by an end user when using end-devices. The quality of a network service, including core network and access network, is determined by "access QoS" and "backbone QoS". Hence, the E2E-QoE is considered as a conjunction of all intermediate aspects: QoP, QoD and QoS. For this reason, QoE is an important factor for an operator to properly design and manage its network.

Regarding another point of view of the relationship between QoS and QoE, we now analyze the *quality management* (QM), an indispensable term for every network operator. QM aims not only to ensure that a network service is consistent but also to achieve consistency. QM includes two main modules: *quality assurance* (QA) and *client satisfaction* (CS) (Figure 2.3). The term QA is defined as systematic activities implemented in a quality system to fulfill all quality requirements. While QA tries to design and meet worthy service specifications, CS metrics seek to capture the degree to which those requirements translate into services/applications that fulfill the user's expectations. The QoS is actually related to the QA and the QoE is related to CS. This relation is explained by the fact that QoS is often part of enforceable SLAs based on specific performance levels while QoE refers to the perceived quality of the user engagement.

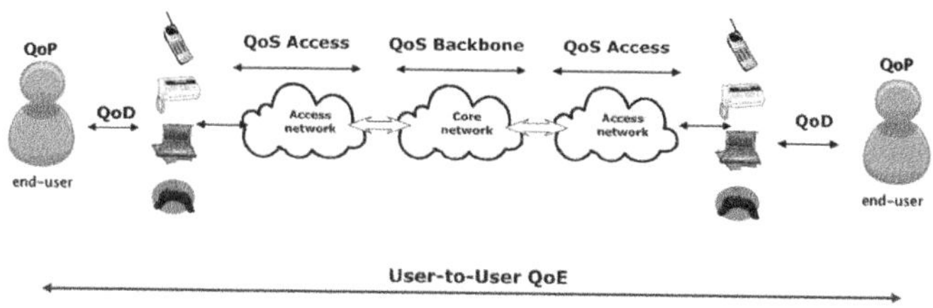

Figure 2.2. *Quality chain in an E2E service*

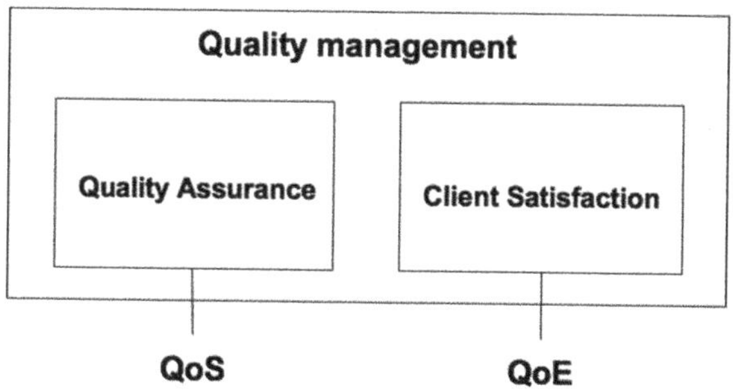

Figure 2.3. *QoE and QoS in quality management module*

According to the qualitative dependent relationship between QoE and QoS, Fiedler *et al.* [FIE 10b] give a schematic relation describing the impact of QoS problems on QoE. In Figure 2.4, the *x*-axis is the QoS disturbance and the *y*-axis is the QoE value, which is described by an MOS score [REC 06] (i.e. a scale of 1 to 5 where 5 is the best and 1 is the worst). As a function of QoS disturbance, the QoE is divided into three areas, separated by *threshold 1* and *threshold 2*. We now analyze each of these areas (Figure 2.4):

– *Area of Constant QoE*: In this area, the QoS is good enough to give a good and constant QoE. A little growth of QoS disturbance does not affect the QoE value. For example, the jitter buffer can eliminate small delays and delay variations, without the recognition of additional delays by the user.

– *Area of decreasing QoE*: When the QoS disturbance exceeds *threshold 1*, the QoE value begins to decrease. Then, the higher the QoS disturbance grows, the higher the QoE value decreases. In cases of high QoE, a little additional QoS disturbance might have a considerable impact on the QoE. On the contrary, for low QoE, the particular additional QoS disturbance might no longer be that critical.

– *Area of unacceptable QoE*: After reaching *threshold 2*, the perceived quality becomes unacceptable. For example, the service might stop working because of timeout problems. The dashed line illustrates that user might quit the service from this point (*threshold 2*). For another example, *threshold 2* represents the cancellation rate of Web browsing when the loading time exceeds a limit.

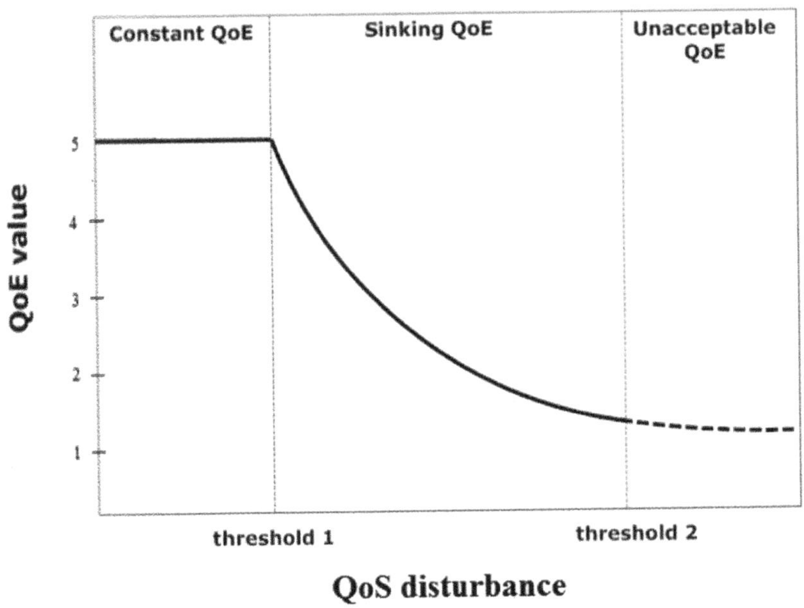

Figure 2.4. *General mapping curve between QoS and QoE*

Concerning the mapping between QoS and QoE in Figure 2.4, the network provider should detect the moment of *threshold 1* and propose solutions to improve service quality as long as the QoE is falling into the *decreasing area*. The goal is not only to solve the problem of quality degradation but also to avoid the churn problem (i.e. users give up the operator). Hence, the QoE is the most important factor for an operator to properly design and manage its network. A good QoS gives a good QoE. However, maintaining all the QoS parameters will not ensure a satisfied user. For example, in a wireless network, if there is no coverage a short distance away, this causes a bad QoE, although there is an excellent throughput in one part of a network.

In order to guarantee a good QoE in network system, apart from understanding the QoE/QoS relationship, we also need to comprehend the impact of networking on QoE. The next section discusses this issue.

2.7. Impact of networking on QoE

According to the impact of networking in QoE, with the continuous development of network services, it is giving new features and advances. It is not easy to explain how networking can affect the QoE because there are many research topics, technologies and applications related with the networking topic. First, based on [MAR 10], we present the layered approach similar to the Transmission Control Protocol (TCP)/IP model to explain this issue. Then, we talk about the impact of some network criteria on the QoE.

2.7.1. *Layered classification of impacts on QoE*

2.7.1.1. *Link layer*

The limit of the maximum speed that is transmitted in a link gives the link layer a fundamental role in the UE. In fact, it decides how the data are transported over the links and how the end users can access a link. In addition, the link layer can prioritize some kind of traffic in adjusting QoS parameters. On the other hand, this layer supports not only wired link technologies but also wireless technologies. Here we cite some examples: Wi-Fi [HEN 02], Wimax [NUA 07] and UMTS [KAA 01] that provide the wireless infrastructure for 3G networks. Besides the shared access technologies we mentioned previously, there are also point-to-point technologies such as digital subscriber line (DSL) [STA 99] and fiber to the home (FTTH) [KOO 06], which can provide QoS. The link layer technologies have the same aim but different performance, capacity and behavior. In addition, the deployed scenario and the external conditions influence their performance. Therefore, the context and scenario description is very important in QoE measurement.

2.7.1.2. *Network layer*

The network layer includes IPs. The IP performance depends on various factors. First, it is the routing path that data packets must follow to reach the destination. The routing path is influenced by three factors: the end points, the connectivity from ISPs to the customers and the relationship between ISPs. Building the routing tables in the routers of ISPs needs Border Gateway Protocol (BGP) [RAM 00] that manages the autonomous systems (ASs). Thus, the performance of the Internet is based on the way the ISPs establish routing policies through BGP. Therefore, many research projects related with IP carry the relationship between ASs and ISPs. Considering the QoE makes these projects more realistic and closer to the results in a real deployment. The other factor affecting IP is mobility. IP was originally designed in considering only wired network where hosts are fixed. However, the appearance of wireless technology as well as laptops, hand-held devices, poses a challenge to IPs. In fact, the QoE perceived by end users changes due to the fact that mobility affects delay, jitter and loss rate.

2.7.1.3. *Transport layer*

The transport layer plays an important role in the QoE perceived by end user. The main protocols of this layer are TCP [AWE 02] and User Datagram Protocol (UDP) [KUR 05]. The connection-oriented protocol TCP process different E2E flow congestion control mechanisms, which affect the QoE [TRI 10]. Besides TCP, there are also other similar protocols such as Stream Control Transmission Protocol (SCTP) [STE 01] or Datagram Congestion Control Protocol (DCCP) [KOH 06]. On the other hand, the connectionless protocol UDP is used for real-time transmission where the flow control is not necessary. Actually, the peer-to-peer traffic that has been growing for years affects the quality of other flows in the Internet because peer-to-peer applications open lots of connections to both UDP and TCP. Furthermore, different flows competing for the available network resource imply the overload issue of network system. According to Martinez-Yelmo *et al.* [MAR 10], the IETF Low Extra Delay Background Transport (LEDBAT) Working Group (WG) carries this problem to develop lower-than-best-effort transport protocols that avoid decreasing the QoE appreciated in the rest of applications running on the Internet.

2.7.1.4. *Application layer*

The application layer is an important layer in terms of QoE because it is the only thing that users can really appreciate and interact with. In fact, the end users do not take into account the technical metrics under their applications. They just consider the subjective media quality of applications they are using. For example, the simplicity of *install and use* is an important feature of a VoIP application. If the latter was able to cross NATs and firewalls without reconfiguring any network equipment and the users could install it in any computer and network without any complex configuration, this application would obtain a good QoE. Skype [CHE 06] is such an application that has had much success due to good QoE with the simplicity of install and use. QoE has different meanings in different kinds of applications. While the sound fidelity and ability to smoothly take turns in a conversation give a good QoE for a VoIP application, a multimedia streaming application with a large and clear video image has a positive QoE. For a Web browser, good QoE means the Web content is retrieved fast enough.

2.7.2. *Impact of user mobility on QoE*

In next-generation wireless networks, the IP-based transport through the radio and core network parts is provided by the all-IP-paradigm solution. However, this concept requires a precise management of user mobility. Therefore, the impacts of user mobility on QoE are an important issue to consider, and studied by Bernardo *et al.*[BER 08] and Ciubotaru and Muntean [CIU 09].

Bernardo *et al.* [BER 08] propose mechanisms in order to evaluate the QoE that users perceive with different QoS-aware mobility management strategies. In their work, the authors use a test bed that can evaluate the QoE in a real heterogeneous wireless access network. By contrast, Ciubotaru *et al.* [CIU 09] propose another novel quality-aware approach (smooth adaptive soft-handover algorithm (SASHA)) to handover on load balancing among different networks using a comprehensive function for decision-making. These approaches are described definitively as follows.

In [BER 08], the authors presented a brief overview of the test bed in the mobility management (MM). This test bed reproduces a beyond 3G heterogeneous radio access network that consists of three radio access networks (RANs): universal terrestrial radio access network (UTRAN), GSM EDGE radio access network (GERAN) and WLAN interfacing a common core network based on DiffServ/MPLS protocols. In this test bed, user equipment is connected to ingress routers (IRs) through traffic switching. The authors present three handover types:

– Horizontal handover where handover is made between base stations of the same RAN, called intra-RAN handover.

– Intra-IR vertical handover where the handover is performed between base stations of different RANs.

– Inter-IR vertical handover where MM plays a crucial role with an IR change. On the core network side, the data information of the user under test (UUT) is encapsulated into MPLS tunnels from the egress router to one of the IRs for downlink and uplink.

After testing different handover types, the QoE of the UUT is depicted by MOS values. The authors used an FR model-based objective metric to obtain a satisfaction-level measurement.

Studies of Ciubotaru and Muntean [CIU 09] focused in a heterogeneous wireless network environment where the mobile device uses different wireless technologies to access to several networks. In order to maximize users' perceived quality in a dynamic environment, a quality-oriented mobility management solution with efficient resource allocation is required. Multimedia mobility management system (M3S) is an application-level framework for delivering high-quality multimedia content to mobile clients. A novel Quality of Multimedia Streaming metric (QMS) is presented to depict and quantify the impact of separated network parameters to the multimedia delivery quality. M3S provides handover management and efficient quality-oriented resource allocation using the SASHA, which performs handover between different networks by transferring the load from one network to another. QMS is described by equation [2.6] with the communication channel i:

$$\begin{aligned} QMS^i = \; & w_1 QoS^i_{grade} + w_2 \times QoE^i_{grade} \\ & + w_3 \times Cost^i_{grade} + w_4 \times Peff^i_{grade} \\ & + w_5 \times UPref^i_{grade} \end{aligned} \qquad [2.6]$$

User preference grade ($UPref^i_{grade}$) is the score of the user that expresses the user's preference for the network interface used by the communication channel i.

User's perceived quality grade (QoE^i_{grade}) is the user's perceived quality that is calculated based on the received content's video quality. QoE grade for each communication channel is determined by distributing the overall QoE according to the channel's corresponding rate share:

$$QoE^i_{grade} = QoE_{overall} \times RateShare^i \qquad [2.7]$$

This algorithm executes a quality-oriented handover by sending the load from one network to another. The handover decision-making algorithm is based on the QMS grades for each communication channel. QMS represents a more comprehensive metric to construct a handover management solution oriented on users' perceived quality by taking several QoS and QoE related parameters into account. This approach aims to maximize the end users' perceived quality (QoE) when streaming multimedia content by efficiently using all the communication resources available. M3S uses the novel algorithm (SASHA) to distribute the load over the communication channels based on their estimated contribution in order to deliver high-quality multimedia content.

The two approaches mentioned above show the influence of user mobility on the QoE. The next section discusses the impact of network resource utilization and management on the QoE.

2.7.3. *Impact of network resource utilization and management on QoE*

Proper management of the network resource is a good way to handle and maintain the system quality, so it also has a significant impact on end user perception. Indeed, there were many systems taking into account this feature such as [YAM 07] and [AGB 08].

In [YAM 07], a network resource management system for resource access control facility (RACF) based on the end user QoE is presented. The authors constructed a concept of a network resource management system (NetRM) to provide service-oriented network services via a Web services interface (WSI). NetRM distributes and supervises the suitable E2E network resources to meet the service request. NetRM provides guaranteed and scheduled network services to network service clients.

In [AGB 08], the authors propose a more concrete approach based on a QoE management methodology, which aims at maximizing the user's QoE while preserving network resources at the same time. A statistical modeling technique, discriminant analysis, has been employed to implement this methodology. Because of the prediction model, network operators can anticipate the user's experience and then allocate network resources accordingly. A statistical technique can be used to predict a qualitative attribute of an individual from known quantitative variables. In order to perform a discriminated analysis, we need a set of observations whose group memberships are already known. The general formula for a discriminated function is given as:

$$f_{km} = u_0 + u_1 X_{1km} + u_2 X_{2km} + \ldots + u_p X_{pkm} \qquad [2.8]$$

where:

f_{km} is the predicted discriminated score for case m in the group k_i,

X_{ikm} is the value of the quantitative predictors X_i for case m in group k,

u_i is coefficients for variable i, for $i = 1..p$.

The results taken from this chapter illustrate how to maximize the utilization of network resources and provide different levels of service quality.

We consider these two approaches [YAM 07, AGB 08] as adaptive systems. They both used resource management functions to maintain required QoE and QoS. The main differences between the two approaches are the functions used. In [YAM 07], in order to maintain the QoE, a network resource management function was implemented in each access network and utilized when the communication service was set up to allocate required bandwidth. Instead, in [AGB 08] a discriminated function was used and a statistical modeling technique was implemented to maximize the user's QoE while preserving network resources.

2.7.4. *Impact of billing and pricing*

We will discuss how billing and pricing can have an impact on QoE. Takahashi *et al.* [TAK 08] studied using a particular application based on IPTV service and focused on QoE requirements in consideration of service billing for this application.

In their work, QoE was defined as how well an IPTV service satisfies the customer's expectation. It was found that the service billing had a strong correlation with the customer's expectation for service quality. The authors gave an example where service has a long channel zapping delay. An IPTV customer subscribing to a basic class accepted this channel zapping time. On the other hand, the premium IPTV customer who pays a more expensive service charge was dissatisfied with the same channel zapping time. Since customers tend to expect a better service quality as the service charge increases, the customer's evaluation for the service strongly depends on its service billing.

As an example of a QoE metric with service billing, the authors propose *end user's utility* (U_s), which is a relative satisfaction level for service quality. Another metric is *service quality metric* (T_s) for IPTV, e.g. channel zapping delay, control command response time, STB startup time and picture quality. The third metric is *service billing* (B_s), i.e. the service charge for a service S. Finally, U_s can be defined as $U_s = f(T_s, B_s)$. In this formula, f is a weight function that transforms the absolute service quality metric, T_s, with service billing as a relative end user utility value U_s. Therefore, even if two customers have the same service quality metrics, their satisfaction levels (U_s) can be different because each customer may pay a different service charge. The weight function is based on the types of services, the types of customers and the policies of service billing.

2.8. Conclusion

Nowadays, emerging network services (e.g. VoIP, video on demand and IPTV) are becoming increasingly numerous. Network service providers try to improve the quality of the offered services perceived by the end users. In this chapter, after presenting QoE/QoS relationship, QoE metrics and some QoE measurements models, many issues that may impact user perception in many different ways are also given, such as the method of measuring and monitoring QoE, modeling of user satisfaction as well as the impact of user mobility and network resource utilization.

3

Content Distribution Network

In this chapter, first, we present the general architecture of a content distribution network (CDN) system with its components and functions. Then, we describe in detail its two main layers: the routing layer and the meta-routing layer.

3.1. Motivation

Nowadays, the demand for content on the Internet is increasing rapidly. This needs an efficient and scalable content distribution solution. Implementing multiple servers placed at geographically distributed locations replaces using single-server schemes. The general idea of this evolution is to bring the servers geographically closer to clients. This new architecture, called CDN, will improve the performance of the system, reduce network load and provide better fault tolerance. Currently, there are many successful CDN providers such as Akamai [AKA 98] and Digital Island [DIG 96].

The general operation mechanism of a CDN service is depicted in Figure 3.1. When a customer (e.g. a Website and VoD service) subscribes to a CDN service, the customer provides data to the CDN provider. The latter will implement a distribution network in using replica server scheme. More precisely, a CDN provider replicates the content data from the origin server of the customer to the replica servers scattered over the Internet. The operation mechanism of a user's request is described step by step as follows. First, the end user sends a request to the Website (step 1). Then, the Website forwards the request to the CDN service (step 2). Finally, the CDN provider uses several mechanisms to select the most appropriate server with which to serve the end user (step 3).

Figure 3.2 shows a detailed architecture of a CDN system [PEN 04, PAT 08]. The CDN focuses on serving both Web content and streaming media. We now present some main components of this architecture: origin server, replica servers, clients, access network, distribution network, content manager and redirector.

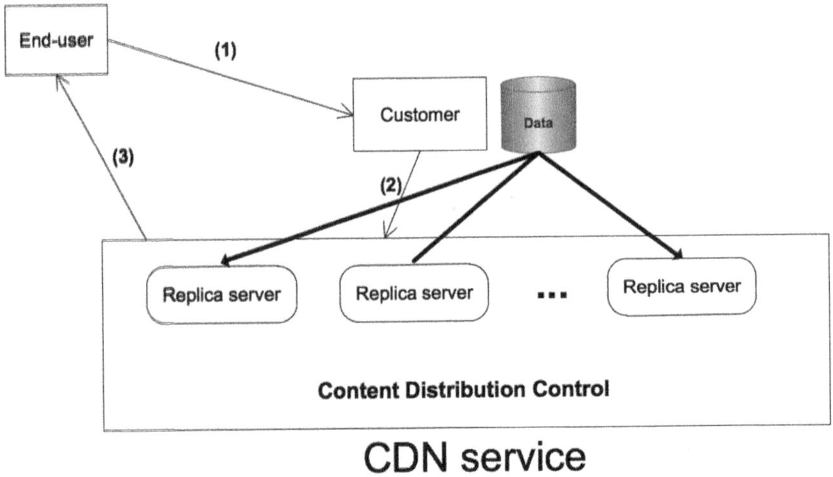

Figure 3.1. *CDN service*

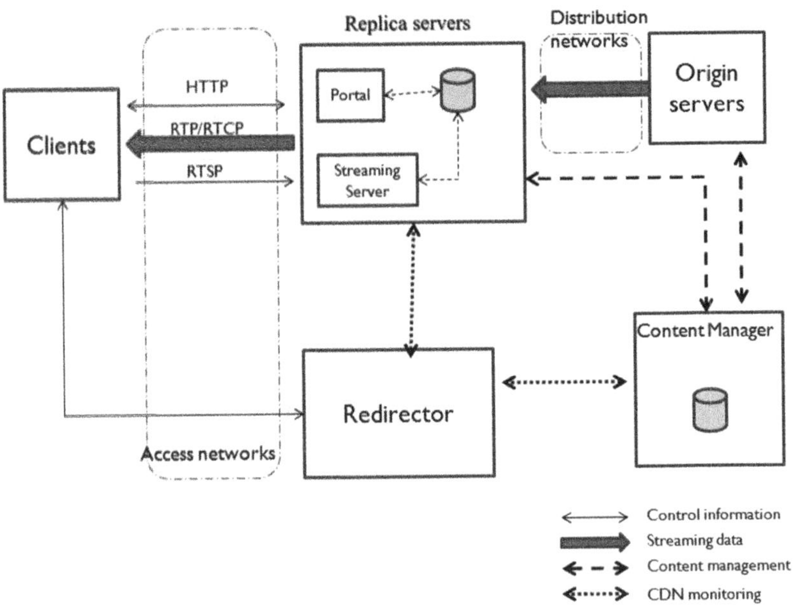

Figure 3.2. *General architecture of CDN system*

An *origin server* contains the information to be distributed or accessed by clients. It distributes the delegated content to the replica servers of the CDN through the *distribution network*. The replica servers act as proxy/cache servers with the ability to store and deliver content. The structure, number and location of replica servers are used to classify the CDN and the selection algorithm executed to determine the server who serves each issued request. *Clients* are represented by PCs or special set-top boxes that request and download a particular piece of content stored somewhere in the CDN. The CDN usually works with clusters of clients rather than individual clients. The clients access the service provided by the CDN through different *access networks*, depending on the ISP.

Distribution network interconnects the origin servers with the multimedia streaming servers (MSSs) to deliver media objects within the streaming media CDN. The content manager controls the media objects stored in each replica server, providing this information to the *redirector* in order to get each client served by the most suitable replica server. The redirector module provides intelligence to the system because it estimates the most adequate replica server for each different request and client.

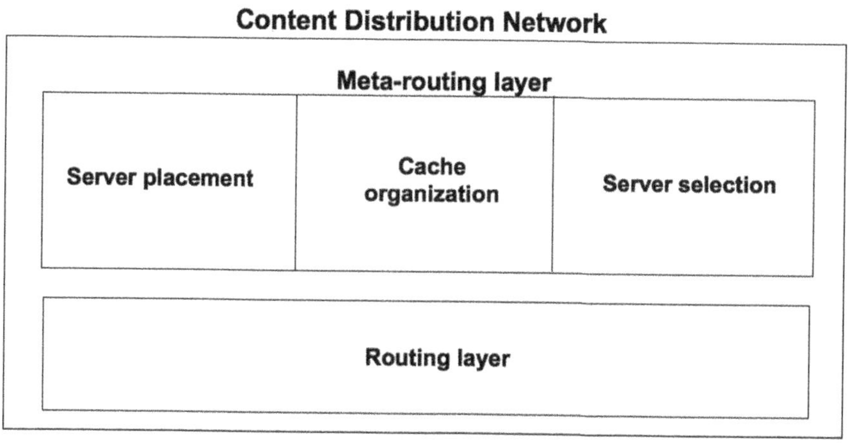

Figure 3.3. *Two layers in a CDN architecture*

A functional CDN architecture includes two key layers: the *routing layer* and *meta-routing layer* (Figure 3.3).

– The routing layer: the core network components (e.g. routers and switches) take into account the routing process in providing the basic network connectivity and guaranteed Quality-of-Service (QoS) for a specific request or traffic. This layer plays no active role in the server selection process.

– The meta-routing layer: CDN providers replicate content to replica servers. This layer takes into account some modules such as server placement and replica server selection.

These two independent layers are the brain of a CDN architecture (Figure 3.3). While the routing layer takes into account the path finding process to optimize data delivery tasks, the meta-routing layer carries out some modules. In this book, we focus on the server selection module and propose a novel server selection policy based on user perception.

Definitively, for each layer, we propose a novel approach:

1) *In the routing layer*: we propose a routing algorithm, called QoE Q-learning-based adaptive routing (QQAR). The latter is based on QoE in using Q-learning algorithm. We consider the user's feedback as reward to formalize our routing model to a reinforcement learning (RL) model. In fact, the goal is to optimize the end user's QoE of the routing system.

2) *In the meta-routing layer*: among several modules in this layer, we focus on the server selection module. We propose a new server selection algorithm in formalizing it to the multi armed bandit (MAB) formalization. In fact, this algorithm not only chooses the most appropriate server that gives the good QoE level, but also takes into consideration the exploitation-exploration trade-off.

3.2. Routing layer

As a layer below the meta-routing layer in a CDN, the routing layer takes into account the routing process to transfer data between end user and selected server. The goal is to choose the best path that gives good user perception on end user side. For traditional routing process in CDN, providers usually use QoS routing systems, which are based on technical parameters and widely use routing protocols. In this section, first, we survey the QoS routing approaches they used in their CDN systems. Then, we present some classical routing algorithms. Finally, we discuss QoS routing with different protocols.

3.2.1. *Routing in telecommunication network*

Routing is one of the main features of the network layer, which is responsible for deciding on which output line an incoming packet must be retransmitted. In general, there is direct delivery, which corresponds to the transfer of a datagram between two computers on the same network, and indirect delivery, which is implemented as at least a router that separates the original sender and final recipient. Indirect delivery is necessary for determining to which router an IP datagram should be sent, according to its final destination. It is possible to use a specific routing table for each router to determine to which route a datagram output should be sent for any network. The main contents of a routing table consist of quadruples (*destination*, *gateway*, *mask*, *interface*).

There are many routing protocols, which can be divided into the following two categories:

– The intra-domain protocols "Interior Gateway Protocol (IGP)" (such as Routing Information Protocol (RIP)), Open Shortest Path First (OSPF), Interior Gateway Routing Protocol (IGRP) and IS-IS) that calculate the routing tables within the same autonomous system (AS).

– The inter-domain protocols "Exterior Gateway Protocol (EGP)" (such as Border Gateway Protocol (BGP)) that calculate the routing tables between nodes belonging to different ASs.

In this section, we focus particularly on the intra-domain protocols that can be divided into two main categories: *distance-vector protocols* and *link-state protocols*.

3.2.2. *Classical routing algorithms*

3.2.2.1. *Distance-vector routing*

For distance-vector routing algorithm (Bellman–Ford routing) [HEI 08], each router has a routing table that indicates each destination, the best known distance and that line to reach it. This algorithm is used by the protocols RIPv1, RIPv2 [ATK 07] and IGRP [SHA 10b]. The estimated distance in the routing table assumes that a metric must be calculated locally for each router (e.g. the hop-count, packet number in the queue, routing time and link cost). The distance-vector routing algorithm can determine the optimal route by exchanging information in the list of metric distances. Furthermore, this information is exchanged between neighbor routers. First, each router starts registering, in its routing table, the links with neighbor routers indicating the line and the estimated distance. Then, periodically, each router sends this to all its neighbor tables extracted from its routing table containing the estimated distance for each destination, thereby updating the routing table. The principle idea of distance-vector routing is to find the shortest path following the imposed metric.

3.2.2.2. *Link-state routing*

With the link-state routing algorithms, each router maintains the same database of network topology. Each router executes the same algorithm in parallel, by building a tree, of which it is the root, to find the shortest path. This tree of the shortest path has the information of the entire path to any destination network, but only the next hop is stored in routing tables. After detecting the change of a link state, the router transmits information via a mechanism of broadcast to all other routers. Like distance-vector routing, the notion of shortest path uses a distance defined by a metric such as hop count and the number of packets in the queue. First, the link-state routing algorithm begins researching the neighbors and assessing the distances to reach them. Then, each estimated distance is sent to all routers in the network. Each router can include in its database the complete network topology, and then build the tree to calculate the shortest path. A typical link-state routing protocol is the OSPF [MOY 08, RÉT 09].

3.2.3. *QoS-based routing*

The user's needs of telecommunications are characterized by an increase in bandwidth requirements (television services such as VoD, Internet access, telecommuting and video telephony). The use of next-generation network (NGN) as part of multimedia applications, services of guaranteed quality, mailing services, mobile services, etc., requires that the delivery service is guaranteed by a QoS control (throughput, delay, jitter, reliability, etc.). Certainly, the above-mentioned routing techniques do not meet these characteristics. Thus, it is necessary to design new routing mechanisms. Routing must take into account the real state of links and routers to perform QoS routing. Taking into account the QoS, routing is organized around three stages: (1) evaluating the state of network links; (2) collecting the information by the routers and (3) determining the routers satisfying these QoS constraints.

However, the authors of [WAN 95] and [MIC 79] proved that a QoS-based routing system met an NP-complete problem in finding a path satisfying more than two non-correlative constraints (e.g. delay, loss rate, jitter and cost) at the same time. Apart from the complexity problem, QoS-based routing faces another issue as follows. In fact, identifying a path for guaranteeing the QoS requirements needs accurate knowledge of the availability of network resources. Most existing routing algorithms assume the availability of precise state information. However, in many cases, this information may be inaccurate and insufficient. Another issue of QoS-based routing is that this kind of routing is no longer sufficient for the new types of content on the Internet: the high-quality multimedia content. Such content is sensitive to end-to-end delay, loss rate, etc., opposed to classic service-level agreements (SLAs) between network operators and service providers.

The approaches of QoS routing can be classified into four categories [HOC 04]:

– those based on the principle of switching (multiprotocol label switching (MPLS));

– those derived from existing protocols (Quality of Service Path First (QOSPF), multipath);

– those based on multiple metrics (widest-shortest path (SW), shortest-widest path (WS));

– those based on learning techniques (Q-learning, Ants, etc.)

We now present the four categories in detail.

3.2.3.1. *Switching-based protocol: MPLS*

MPLS [BOC 10] is a protocol that imposes a fixed route onto different flows to arrive at their destinations. It is based on the concept of label switching. Traffic characterization defining the required QoS is associated with each flow. It is not strictly a protocol of layer 3 (network layer), but rather an intermediary between layer 2 (link layer) and layer 3 (network layer) of the OSI model. MPLS is a technology that offers circuit mode to an IP, to the image of X25 or ATM. The advantages of MPLS can be summarized as follows:

– fast switching at the level of core equipment;

– ability to associate specific routing policies with certain flows.

The idea of label switching is to replace the search for the longest match between the destination address of IP packets, and present prefixes in the routing tables by inserting a label of fixed-length header between the network and link header of the packets. The determination of the next one is then made through the label. This solution is more advantageous, since it no longer relies on finding the longest match of prefixes of variable length, but on finding a fixed-length label.

Prospects for MPLS to improve the QoSs are numerous. Cisco contributed to the standardization of MPLS by offering the tag switching, which uses 3 bits in the *Service Type* field of the IPv4 header of incoming IP packets and places them in the field *Class of Service (CoS)* of the MPLS label. In the MPLS domain, the routing can be performed so as to favor the flow having the highest CoS. Moreover, one of the most interesting aspects of MPLS is the ability to build explicit end-to-end routing, the label-switched paths (LSP). These paths can be determined to meet the QoS constraints. Thus, multiple parallel paths can exist between two routers in the MPLS domain. Each path offers unique characteristics in terms of delay or bandwidth.

3.2.3.2. *Protocols derived from existing protocols*

We now survey some protocols that are derived from existing protocols. The first protocol is QOSPF [ZHA 07], an extension of OSPF. Combined with a reservation protocol, this routing protocol can inform all routers of the capacity of links to support the constraints of QoS. QOSPF is based on the use of two messages: the link resource advertisement (RES-LSA) message and the resource reservation advertisement (RRA) message. The purpose of the RES-LSA messages is to inform all the routers of the link state. This information is used to determine paths. This message type is used when a new router is added to the network or when the load of a link varies. The role of the RRA message is to indicate the resources used by a stream (or rather reserved for this flow). Each router is notified of the resources and the route taken by each flow. The number of RRA messages can become very high if the number of flows increases, which poses a problem of scale.

Classical routing algorithms use a single route to route packets from their source to their destination. To make best use of network resources, we should conduct a multipath routing [HE 08]. Indeed, the routing algorithms can find multiple paths with the same minimal cost. The choice of one of these paths is currently arbitrary. The techniques, namely equal cost multipath (ECMP) [GOJ 03], have been proposed in order to evenly distribute the traffic on these paths. Other techniques suggest an uneven distribution on different paths (optimized multipath (OMP)). There are also extensions to OSPF to perform routing on several paths. OSPF-ECMP can divide the traffic equally on optimal paths. Since these routing techniques are not sensitive to the load, there is no risk of oscillations.

3.2.3.3. *Protocols based on multiple metrics*

Metric combination is used to combine different metrics into a single metric and calculate the path that minimizes the resulting metric. Metric composition can use linear, nonlinear and Lagrange relaxation composition. The key idea in QoS routing is to find a path satisfying multiple constraints on routing metrics like desired bandwidth, delay time and end-to-end QoS. Nowadays, many studies have been introduced to integrate the QoS requirements which are problematic for the routing algorithm. However, only few proposed routing algorithms are suitable for today's Internet such as QOSPF traffic engineering (TE) and the Wang Crowcroft algorithm (WCA). The Enhanced Interior Gateway Routing Protocol (EIGRP) [LEM 10] is also one of the multimetric routing algorithms.

In [GUE 99], the authors proposed the Guerin–Orda algorithm to solve both the bandwidth-constrained and delay-constrained problem with imprecise network states. The goal of the bandwidth-constrained routing is to find the path with the highest probability of accommodating a connection having a precise bandwidth requirement. The goal of the delay-constrained routing is to find a path having the highest probability of satisfying a given end-to-end delay. In fact, it is NP-hard to find the path satisfying these constraints at the same time. The idea of this algorithm is to transform a global constraint into local constraints. More precisely, the approach is to split the end-to-end delay constraint of intermediate links in such a way that all links in the path have the same probability of satisfying the local constraint.

Ma and Steenkiste [MA 97] presented a novel approach to solve the constraints on delay, jitter and buffer space in rate-based scheduling networks. In [WAN 96], the authors proposed a hop-by-hop distributed routing system. Every router predetermines a forwarding neighbor for every possible destination. This approach uses link-state protocol to maintain a complete global state, which is used to determine the shortest-widest path (i.e. the path that has maximum bandwidth and smallest delay).

3.2.3.4. *Protocols based on learning techniques*

In order to design adaptive algorithms for dynamic network routing problems, the researchers used different learning techniques. For RL, technique is defined as learning a policy, a converting of observations into actions, based on feedback from the environment. In RL, the network represents the environment whose state is determined by the number and relative position of nodes, the status of links between them and the dynamics of packets. Each node is considered as an agent who has a choice of actions. Another learning technique is based on the behavior of various biological entities such as an ant/bee colony [MEL 11a, BIT 10a]. Ant colony optimization (ACO) [DOR 06] is influenced by the foraging behavior of some ant species. These ants leave a pheromone on the ground to give information about a favorable path that should be followed by other members of the colony. ACO is widely used in routing systems. Di Caro *et al.* [DIC 05] proposed AntHocNet, an algorithm for routing in mobile *ad hoc* networks. This hybrid algorithm combines reactive route setup with proactive route probing, maintenance and improvement. At the start of a data session, it sets up paths when they are needed instead of maintaining routes to all destinations. Backward ants return to the source to set up paths in the form of pheromone tables indicating their quality as well.

Besides these techniques, there are also various routing algorithms based on a neural network for finding the optimal path [PAT 12, SCH 09]. In [SCH 09], the authors propose a hybrid intelligent method for routing combining Hopfield neural networks (HNNs) and simulated annealing (SA). The proposed method is considered as a modified version of the discrete time equation used to determine the new neuron input. The novel version of the equation improved the HNN convergence and decreased the computation cost. In addition, the SA algorithm is used to obtain the optimal parameters of the HNN. Gelenbe and Lent [GEL 04] propose cognitive packet networks (CPNs) that are based on random neural networks. These are store-and-forward packet networks in which the intelligence is constructed into the packets rather than at the routers or in the high-level protocols. CPN is a reliable packet network infrastructure that incorporates packet loss and delays directly into user QoS criteria and uses these criteria to do routing. The major drawback of algorithms based on cognitive packet networks is the convergence time, which is affected when the network is heavily loaded.

The Q-routing approach was first proposed by Boyan and Littman [BOY 94]. This is a technique where each node makes its routing decision based on the local routing information, represented as a table of Q-values that estimate the quality of the alternative routes. These values are updated each time the node sends a packet to one of its neighbors. However, when a Q-value is not updated for a long time, it does not necessarily reflect the current state of the network and hence a routing decision that is based on such an unreliable Q-value will not be accurate. The update rule in Q-routing does not take into account the reliability of the estimated or the updated Q-value because it depends on the traffic pattern and load levels. In fact, most of the Q-values in the network are unreliable.

Mellouk *et al.* [?] propose a continuous state-dependent routing algorithm based on a multipath routing approach exploiting the Q-routing algorithm. The global learning algorithm finds K best paths in terms of cumulative link cost and optimizes the residual bandwidth and the average delivery time on these paths. The technique used to estimate the residual bandwidth and the end-to-end delay is based on the RL approach to take into account dynamic changes in networks. The proposed system, called bandwidth delay K optimal paths Q-routing algorithms (BDKOQRA), takes into account static and dynamic QoS criteria and is based on inductive approaches with continuous learning of dynamic network parameters such as link capacity, end-to-end delay and residual bandwidth. A load balancing policy that depends on the traffic dynamic path probability distribution function is also supported.

In the last few decades, many research areas have been massively influenced by the behavior of various biological entities and phenomena. Hence, there are also routing algorithms inspired by bee life [BIT 10b]. Teodorovic and Dell'Orco proposed bee colony optimization (BCO) [TEO 05], which is an improvement and generalization of the bee system. This proposal is able to solve combinatorial problems characterized by uncertainty. The authors used a fuzzy bee system representing one of the possible BCOs on the ride-sharing problem. Inspired by BCO, Markovic *et al.* [MAR 07a] tailored BCO-RWA for the routing and wavelength assignment (RWA) problem in all-optical networks regardless of wavelength conversion in intermediate nodes. This proposal has been applied to a static case in which lightpath requests are known in advance. Authors proved that BCO-RWA can produce optimal or near-optimal solutions in a reasonable amount of computer time.

So, we can see that all approaches mentioned above have taken into account only the QoS criteria (technical parameters) in ignoring the QoE notion. This lack motivates us to integrate the QoE into a routing system. This work is described in Chapter 4. We work not only on the routing layer but also on the meta-routing layer, which is presented in the next section.

3.3. Meta-routing layer

The basic idea of a CDN is to move the content from the locations of origin servers to locations closer to end users, at the edge of the network. Apart from routing mechanisms in core networks (that is realized by routing layer modules), we explain in this section how the CDN provider places the replica servers, how they build caching policy and how they select the appropriate replica server to serve the user request. Among various modules of meta-routing layer, we survey in this section three main modules (Table 3.1): *Server placement*, *cache organization* and *server selection*. Then, we explain why we have chosen the server selection module among the three modules. Finally, we focus on the "server selection" module in proposing a new selection algorithm, which will be presented in Chapter 5.

Modules	Approaches
Server placement	Minimum K-center problem [JAM 00] Facility location problem [QIU 01] k-hierarchically well-separated trees [BAR 96] Topology-informed placement strategy [JAM 01]
Cache organization	**Caching techniques:** [PAT 08] Query-based scheme Digest-based scheme Directory-based scheme Hashing-based scheme **Cache update techniques:** [PAT 08] Periodic update Update propagation On-demand update Invalidation
Server selection	**Algorithms:** Round-Robin [SZY 03] Globule project [PIE 06] [KAR 99] [AND 02] [ARD 01] **Mechanisms:** Client multiplexing [BHA 97, BEC 98] HTTP redirection DNS indirection Anycasting Peer-to-peer

Table 3.1. *Approaches for meta-routing modules*

3.3.1. *Server placement*

In order to minimize the average content access latency perceived by end users and the network bandwidth consumption, CDN needs a server placement module to decide where on the network to place the servers, which store replicas of data. Intuitively, replica servers should be placed closer to the clients to reduce latency and bandwidth consumption. In this section, we give a survey of the approaches concerning server placement issues.

To resolve the server placement problem, there are some theoretical approaches such as *minimum K-center problem* [JAM 00], the *facility location problem* [QIU 01] or *k-hierarchically well-separated trees* [BAR 96]. For the *minimum K-center problem*, the goal is to minimize the maximum distance between a node and the nearest center with a given number of centers. Similar to the latter, the *facility location problem* has a given constraint that is the total cost in terms of building facilities and service clients. The third issue, *k-hierarchically well-separated trees*, is based on graph theory.

These theoretical approaches described above are not easy to implement in real CDN systems because they are computationally expensive and they do not consider the characteristics of the network. Therefore, some heuristic solutions have been developed. Qiu *et al.* [QIU 01] propose a server placement approach based on a greedy algorithm [KRI 00]. The goal of this algorithm is to choose M servers among N potential sites. One site is chosen at a time. First, to determine the suitability for hosting a server, each of the N sites is evaluated individually. The site having the lowest cost (e.g. bandwidth consumption) is chosen. Then, the iteration continues until M servers have been chosen. This algorithm assumes that users direct their accesses to the server which has the lowest cost. The *l-backtracking greedy algorithm* is discussed in [JAM 01]. The difference from the basic, algorithm is that in each iteration, the *l-backtracking greedy algorithm* verifies all the possible combinations by deleting l of the already placed servers and replacing them with $l + 1$ new servers.

Jamin *et al.* [JAM 01] propose a *topology-informed placement strategy*. The authors assume that the nodes with the highest outdegrees[1] can reach more nodes with smaller latency. They place servers on candidate hosts in descending order of outdegrees. However, as one has no detailed information of network topology, each node represents a single AS. Each AS-level BGP peering corresponds to a node link.

The experiments of these above-mentioned strategies show that *the topology-informed placement strategy* can perform almost as well as *the greedy placement*. They show that using router-level topology information gives better performance than exploiting only AS-level topology. In addition, they proved that the performance improvement degrades with the increasing server number.

After placing servers in the network, another important task of a CDN service provider is to organize the server cache. This work is clarified in the next section.

1 The node outdegree is the number of other nodes it is connected to.

3.3.2. *Cache organization*

We now present the cache organization of a CDN system. *Cache organization* includes caching techniques and cache update techniques. The goal is to guarantee the availability, freshness and reliability of content. The performance of a CDN system will be improved if the cache is well organized. In this section, we present cache organization in two parts: caching techniques and cache update techniques [PAT 08].

The content caching can be classified into two classes: *intra-cluster* and *inter-cluster* caching. The intra-cluster involves different schemes such as the *query-based scheme*, *digest-based scheme*, *directory-based scheme* and *hashing-based scheme*. In the query-based scheme [WES 97], CDN server broadcasts a query to other cooperating CDN servers for each cache miss. The CDN server has to wait for the last "miss" reply of their neighbors to conclude that there is no peer who has the requested content. Hence, the problems of this scheme are the significant query traffic and the delay. The digest-based scheme [ROU 98] focuses on the problem of the flooding queries in the query-based scheme. In the digest-based scheme, each CDN server maintains a digest of content stored by other cooperating replica servers. The updating CDN server informs the cooperating replica servers of content updates. Based on the content digest, a CDN server can decide which particular replica server should be used to route a content request. However, due to the frequent exchange of update messages to ensure having the correct information about each other, the update traffic overhead is a drawback of this scheme. The directory-based scheme [GAD 97] is introduced as a centralized version of the digest-based scheme. In this scheme, the centralized server has the information of all replica servers in a cluster. The CDN server only informs the directory server when local updates occur. If there is a local cache miss, CDN server will query the directory server. The directory server receives all the information from cooperating replica servers. The problems of this centralized scheme are potential bottlenecks and single points of failure. In the hashing-based scheme [KAR 99, VAL 98], all servers use the same hashing functions. The designated server keeps the content based on IP addresses of the CDN servers and the hashing function. This designated server carries out all requests for that particular content. Because of the smallest implementation overhead and highest content sharing efficiency, the hashing-based scheme is more efficient than other schemes.

The inter-cluster cooperative caching cannot apply the hashing-based scheme because representative servers of different clusters are normally distributed geographically. There is only the query-based scheme that can be applied to inter-cluster caching. A cluster will query its neighbor cluster if it fails to serve a content request. The neighbor will reply a hit message if it finds the requested content. On the contrary, if the neighbor does not find the requested content, it will forward this request to an other neighbor. The hashing-based scheme is used by all the servers of a cluster. A server only queries the designated server of that cluster to serve a content request.

To ensure the freshness of content, a CDN provider should take into account *the cache update mechanism.* Clearly, with the delays involved in propagating the content, the content may be inconsistent or expired. Hence, the CDN provider must address the problems about how long different content is to be considered fresh, when to check back with the origin server for updated content, etc. We now discuss some cache update techniques: *periodic update, update propagation, on-demand update* and *invalidation.*

In the *periodic update* technique, caches are updated regularly. The drawback of this approach is that significant levels of unnecessary traffic are generated from update traffic at each interval. With the *update propagation* technique, the active content is performed and pushed to the CDN cache servers. So, the updated version of a document is sent to all caches even if there is a change of document at the origin server. Due to the frequent changing of content, this approach creates excess update traffic. According to *on-demand update*, one propagates the latest copy of a document to the replica server based on the prior request for that content. This technique assumes that the content is not updated unless it is requested. The drawback of this technique is the back-and-forth traffic between the cache and origin server. In the *invalidation* technique, the invalidation message is broadcasted to all surrogate caches when a document is modified at the origin server. While the document is being modified, the surrogate caches are blocked. After that, the cache has to fetch an updated version of the document. The disadvantage of this technique is that it does not make full use of the distribution network for content delivery.

Because of the cache organization, CDNs help the content provider to ensure a freshness content and the consistency of CDN sites. In fact, the content providers themselves construct the policies and heuristics for caching policies. Then, the content providers give the caching policies to the CDN provider, which it will deploy to its caches. Based on these rules, the caches know how to maintain the freshness of content through ensuring consistency.

The tasks of server placement and cache organization are considered as preparation-related process of a CDN service. The first one, the server placement module, is used to place the replica servers in order to minimize the average content access latency and the network bandwidth consumption. The second one, the cache organization module, takes into account the cache management process to guarantee the availability, freshness and reliability of content. Despite this, these two modules are launched in the preparation phase where one designs the infrastructure for a CDN service. In this book, we focus on the module that appears in the main phase of a CDN: the server selection module.

3.3.3. *Server selection*

As explained earlier, we focus on the server selection module among the modules in the meta-routing layer. The server selection module is used to direct the user requests to the suitable replica server. It could direct the request to the closest replica server. However, the closest server may not be the best one for servicing this request. So, apart from hop-count criteria, they should consider other metrics such as network proximity, client perceived latency, distance and replica server load. In this section, we survey several server selection algorithms and mechanisms.

3.3.3.1. *Server selection algorithms*

Server selection algorithms are classified into two types [PAT 08]: *adaptive* and *non-adaptive*. The difference between these two types is the consideration of the current condition. In fact, while the adaptive algorithm considers the current condition of the environment to select the replica server, the non-adaptive algorithm only uses some heuristics to select the replica server.

The most common non-adaptive server selection algorithm is *round-robin* [SZY 03], which balances the load between the servers by sending all requests to replica servers in a round. This simple algorithm is efficient for clusters, where all the replica servers are installed in the same place. However, the round-robin server selection algorithm is not suitable for wide-area distributed systems. Another non-adaptive algorithm assumes that the replica server load and the client–server distance are the most influential factors. This algorithm takes into account the client–server distance to direct requests to the replica servers in such a way that load is balanced among them. The Cisco DistributedDirector [DEL 99] applied some non-adaptive server selection algorithms. They also consider the percentage of received client requests and assume that the server receiving more requests is more powerful. They support random request distribution that randomly chooses the replica server. In addition, there are other non-adaptive algorithms that direct requests to the closest replica.

According to adaptive server selection algorithms, we now survey some representative proposals. The Globule project [PIE 06] proposes an adaptive server selection algorithm that always chooses the server closest to the users in terms of network proximity. They use path length as an estimation metric that is updated periodically. This algorithm seems inefficient (that is proven in [HUF 02]) because the metric used is passive and it does not introduce any additional traffic to the network. In another proposal, Karger *et al.* [KAR 99] propose an algorithm to adapt to hotspots. They utilize URL content to generate a hashing function from various identifiers. They use this hashing function to route the user requests to a logical ring involving cache servers with IDs from the same space. This algorithm directs user requests to the cache server holding the smallest ID. In two proposals [AND 02, ARD 01], the authors use user–server latency as an evaluation metric.

They use client access logs and passive server-side latency to measure the client–server latency. The replica server having the minimal latency is chosen. The drawback of this approach is that it needs the maintenance of central database measurements, which restricts the scalability of the system. In another approach, Cisco DistributedDirector [DEL 99] proposes an adaptive server selection algorithm, which carries out a weighted combination of three metrics: inter-AS distance, intra-AS distance and end-to-end latency. Using three metrics makes this approach flexible. However, the deployment for metric measurement is complex and costly. In addition, the active measurement technique gives more additional traffic to the system. Akamai [AKA 98, DIL 02] propose an adaptive algorithm to avoid the flash crowds problem. Their metrics used are: replica server load, the reliability of loads between the client and each replica server and also the available bandwidth of a server.

3.3.3.2. *Server selection mechanisms*

There are many techniques to direct the user requests to the appropriate server among various replica servers. These techniques are classified into five categories [PEN 04]:

1) *Client multiplexing*: for this mechanism, the client holds an address list of candidate replica servers. The client uses this list to choose the replica server to send the requests. Definitively, the DNS server sends the IP addresses of servers to clients. Afterwards, the DNS resolver at the client's side chooses a server among those based on evaluation metrics such as response time. To obtain these metrics, the DNS resolver makes probes to the servers and also collects reports from the clients on the performance of past accesses to these servers. Two approaches [BHA 97, BEC 98] have applied this mechanism. The advantage of this mechanism is that there is no additional communication of request processing. However, there are the following drawbacks. First, the client has to use a customized DNS resolver. So, their software (e.g. Web browser and proxy) must be modified in such a way that the resolver relies on client performance. Second, the DNS infrastructure is dependent on DNS response caching. Consequently, one cannot dynamically change replica servers at the risk of using stale replica server sets. Moreover, reducing the caching time increases the DNS queries, which causes the bottleneck issue for the DNS infrastructure.

2) *HTTP redirection*: this mechanism is very simple and not efficient for a server selection technique. Generally, all client requests are sent to the origin server. Then, the origin server redirects the browser to a new URL at the HTTP protocol level. In this way, the origin server is the only node that carries out redirecting requests. So, it could become a bottleneck.

3) *DNS indirection*: the main difference between *DNS indirection* and the *client multiplexing* mechanism is that while in the *client multiplexing* mechanism, choosing a replica server occurs on the client side, the *DNS indirection* makes the server selection at the DNS infrastructure. Generally, some domain name server implementations

allow the DNS server to assign a host domain name to a set of IP addresses and choose one of them for every client request. With this mechanism, the replica server infrastructure can be easily changed without any danger of using out-of-date information.

4) *Anycasting*: for this mechanism, an anycast address (an IP anycast address or a URL of content) is a notion that defines a group of servers providing the same service. The client adds an anycast address in the destination address field of packets and sends them to the server group. Then, anycast-aware routers direct the packet to at least one of the servers in this anycast group. This anycast-based routing can be implemented into the existing Internet routing infrastructure.

5) *Peer-to-peer*: a peer-to-peer system is today a broad trend on the Internet for disseminating content. Each affiliation in a peer-to-peer system has participant nodes. The affiliations constitute an *ad hoc* network. Since the network changes continuously, each node cannot have the complete information in time about the entire network. The problem of how to efficiently select the best server in a distributed manner without incurring a high overhead of propagating the routing information actually becomes a trend for network operator.

3.4. Conclusion

This chapter has surveyed the general CDN architecture with its two principal layers: the routing layer and the meta-routing layer. In fact, over the last few years, CDN has actually become an efficient solution for the rapid growth and evolution of the Internet. The mechanism of replicating content from the origin server to replica servers can present short access delay and consume less network bandwidth of the Internet infrastructure. In fact, serving content from a local replica server typically has better performance (lower access latency, higher transfer rate) than from the origin server. This chapter plays an important role in clarifying our way of integrating the QoE model into the CDN architecture. This is also the basic premise to explain the need of applying the QoE model to a CDN.

4

User-Driven Routing Algorithm Application for CDN Flow

In this chapter, we will present a new QoE-based routing algorithm, called QoE QLearning-based Adaptive Routing, which is based on a bio-inspired mechanism and uses the Q-learning approach, an algorithm of reinforcement learning. First, this chapter will briefly present the knowledge base of the Reinforcement Learning and Q-routing. Subsequently, the chapter will focus on the routing algorithm proposal.

4.1. Introduction

Over the past few decades Quality of Service (QoS) played an important role in the QoS framework of the Internet. The QoS routing strategy determines the suitable paths for different types of data traffic sent by various applications. The goal is to maximize the utilization of network resources while satisfying the traffic requirements in the network. To achieve this objective, it is necessary to develop a routing algorithm that determines multi-constrained paths taking into account the network state and the traffic needs (e.g. delay, loss rate and bandwidth). In fact, the core of any QoS routing strategy is the path computation algorithm. The algorithm has to select several alternative paths, which can satisfy a set of constraints like end-to-end delay and bandwidth requirements. However, the algorithms to solve such a problem are generally met with high computational complexity. In fact, Wang *et al.* [WAN 96] proved that an end-to-end QoS model with more than two non-correlated criteria is NP-completeness. Apart from the problem of complexity, current QoS routing strategies also have another problem: the lack of consideration of end user perception. Actually, the current QoS routing strategy just focuses on technical parameters in the network by ignoring the end user perception. Unfortunately, from time to time end user satisfaction plays the key role in the competition of network operators. This challenge has devised a new human-centered metric, called quality of experience (QoE).

A QoE-based network system is a promising new notion in which service providers are aware of user perception and can consequently adapt to the dynamic environment to obtain acceptable and predictable QoE. End-to-end QoE refers to the overall satisfaction and acceptability of an application or service as perceived by the end user. End-to-end QoE expresses a subjective measure of users' experiences and becomes a major issue for telecommunication companies because their businesses are highly dependent on customer satisfaction. The average revenue per user can only be increased by value-added services taking into account the complete end-to-end system effects. As an important measure of the end-to-end performance at the service level from the user's perspective, the QoE is an important metric for the design of systems and engineering processes. With the QoE paradigm, we can reach a better solution and prevent the NP-completeness problem due to the goal of maintaining only a QoE criterion instead of optimizing multiple QoS criteria. Furthermore, QoE is really a contribution to real-adaptive control network components in the future Internet.

Such advantages of QoE-based network system motivate us to devise a new kind of routing algorithm for the CDN architecture: the QoE-based routing algorithm. Definitively, in the CDN context, after receiving a user's request, the server selection module is used to choose the most appropriate server by using a selection algorithm (Figure 4.1). Then, the IP address of the chosen server is attached to the request, which is sent down to the routing layer. The latter will take into account the routing process by using a routing strategy. Our proposed routing strategy adapts to the network environment changes to improve user perception. It is a solution that reacts in real time to maintain a good QoE level. The QoE-based routing system does not wait until the client leaves the service. In contrast, it reacts quickly after detecting quality degradation to improve the QoE value of the network system. In addition, this new kind of routing can resolve problems of the QoS routing framework. First, the QoE-based routing strategy is able to avoid the NP-complete problem. It optimizes only the QoE metric instead of optimizing multiple QoS criteria due to the fact that satisfying end users is the most important goal of network operators. Users disregard the technical parameters in the core network when using services. All they care about is service quality they perceive. Second, QoE-based routing can also avoid the churn problem (i.e. the client leaves a service after not being satisfied). In fact, clients will never leave the service if their experience is put as the top priority.

The scalability of a CDN predicts fluctuations of the environment in the core network. In order to react and adapt to changes of network environment, routing strategy needs an efficient learning method. There are lots of learning methods, such as supervised learning (e.g. neural network), that were applied to various routing approaches. In this book, we have chosen the RL, which is defined as a learning policy, a mapping of observations into actions, based on feedback from the environment. RL is considered as a natural framework for the development of adaptive strategies by trial and error in the process of interaction with the environment. Such method is appropriate for a routing approach.

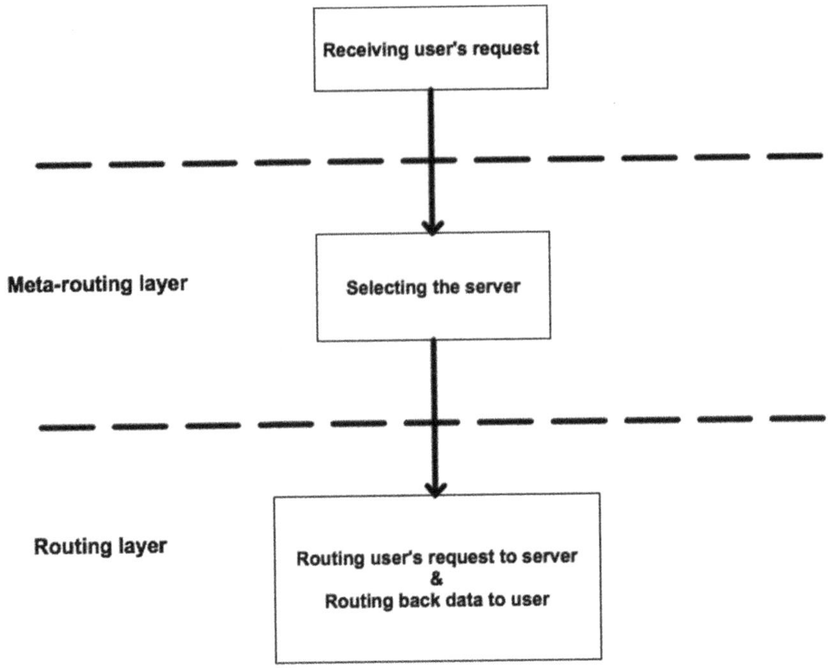

Figure 4.1. *CDN processes*

In this section, we present our routing algorithm, called QQAR protocol. First, we survey the reinforcement learning theory. Then, we describe the QQAR algorithm by mapping the RL model onto our routing model.

4.2. Reinforcement learning and Q-routing

As mentioned previously, our routing algorithm is based on RL that takes into consideration the QoE as environment feedback. Before focusing on the proposal, we present the fundamentals of RL as well as the Q-learning algorithm.

The basis of RL is the *Markov Decision Process* (MDP), which provides a mathematical framework for modeling decision-making. MDP is a discrete time stochastic control process. The MDP model can be considered as a Markov chain to which we add a decision component. Like the rest of its family, it is among others used in artificial intelligence for control of complex systems such as intelligent agents. If the probabilities or rewards in an MDP are unknown, the problem becomes RL [SUT 98]. The latter is a learning method for which only a qualitative measure of error is available. In this case, the agent receives rewards from its environment and reacts to it by choosing an appropriate action for its behavior. Its reaction is then judged against a predefined objective called "reward". The agent receives this "reward" and learns to modify its future actions and achieve an optimal behavior. Action leading to a negative rating will be used less than a positive action.

The RL results from two simple principles:

– For a given state, if an action immediately causes something bad, then the system learns not to take that action when it is in this state in the future.

– For a given state, if all possible actions lead to something bad, then the system will learn to avoid being in this state.

The RL is a learning method for an autonomous agent behavior to adapt to its environment. The goal is to achieve objectives without any external intervention. Figure 4.2 illustrates the principle of RL.

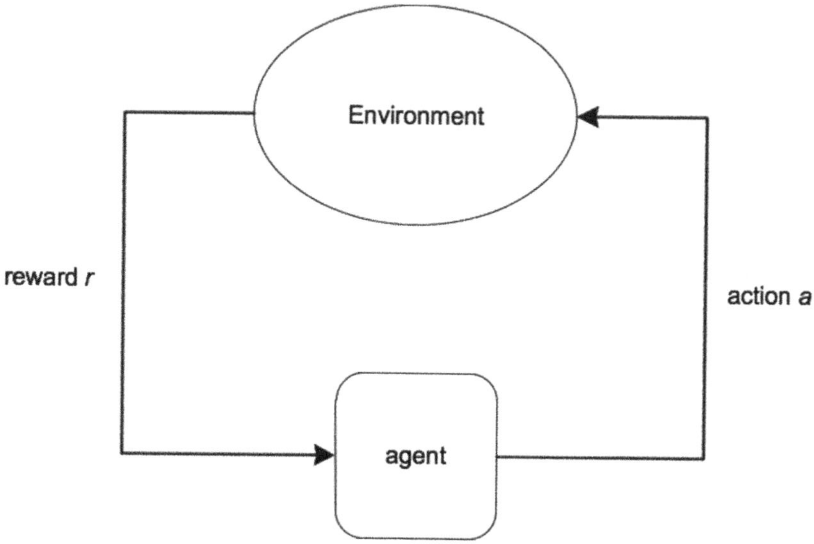

Figure 4.2. *Reinforcement learning model*

In an RL model [KAE 96, SUT 98], an agent is connected to its environment via perception and action (Figure 4.2). Whenever the agent receives, as input, the indication of current environment state, the agent then chooses an action, a, to generate as output. The action changes the state of the environment, and the value of this state transition is communicated to the agent through a scalar reinforcement signal, r. The agent chooses actions that tend to increase by systematic trial and occurrences, guided by any algorithm.

Formally, based on the MDP, the RL model consists of:

– a discrete set of environment states: $\mathcal{S}$;

– a discrete set of agent actions: $\mathcal{A}$;

– a set of scalar reinforcement signals: $\mathcal{R}$.

So the agent's goal is to find a policy, π, mapping states to actions, that maximizes some long-run measure of reinforcement. In RL paradigm, we use two types of value functions to estimate how good it is for the agent to be in a given state:

– State-value function $V^{\pi}(s)$: expected return when starting in s and following π thereafter:

$$V^{\pi}(s) = E_{\pi}\{R_t | s_t = s\} \quad [4.1]$$

– Action-value function $Q^{\pi}(s,a)$: expected return starting from s, taking the action a and thereafter following policy π:

$$Q^{\pi}(s,a) = E_{\pi}\{R_t | s_t = s, a_t = a\} \quad [4.2]$$

where t is any time step and R is the reward.

Regarding the policy iteration notion, the reason for calculating the value function for a policy is to find better policies. Once a policy, π, has been improved using V^{π} to yield a better policy, π', we can then calculate $V^{\pi'}$ and improve it again to yield an even better policy, π''. We can thus obtain a sequence of monotonically improving policies and value functions:

$$\pi_0 \xrightarrow{E} V^{\pi_0} \xrightarrow{I} \pi_1 \xrightarrow{E} V^{\pi_1} \xrightarrow{I} \pi_2 \xrightarrow{E} \ldots \xrightarrow{I} \pi^* \xrightarrow{E} V^* \quad [4.3]$$

where E denotes a policy evaluation and I denotes a policy improvement.

The goal is to find an optimal policy that maps states to actions so as to maximize cumulative reward, which is equivalent to determining the following Q function (Bellman equation):

$$Q^*(s,a) = r(s,a) + \gamma \sum_{s'} P(s'|s,a) max_{a'} Q^*(s',a') \quad [4.4]$$

where γ is the discount rate. The following section presents the mathematical model of RL in detail.

4.2.1. *Mathematical model of reinforcement learning*

We take an example of a model with three states (Figure 4.3). Each state can execute two possible actions a_1 and a_2. The transition from one state to another is performed according to the probability P_{21}^{a1} and P_{21}^{a2}, leading to a reward signal respectively r_{23}^{a1} and r_{23}^{a2}.

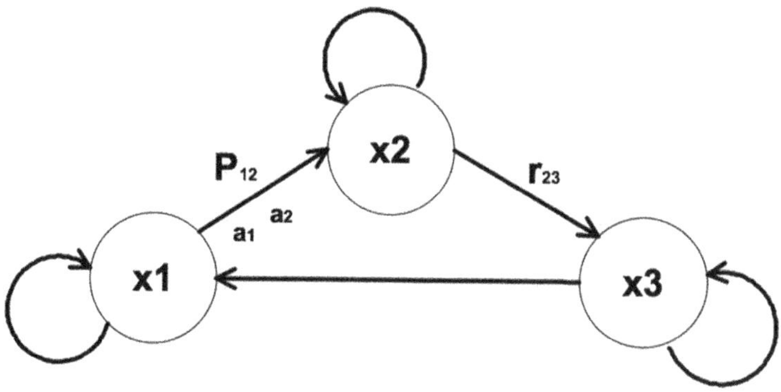

Figure 4.3. *Example of a Markov process with three states and two actions*

These are some notations:

– X: the state set of the system and $x_t \in X$ is a state of the system at time t.

– A_x: the set of possible actions of state x.

– A: set of all actions.

– $a_t \in A_x$: applied action in time t of state x_t.

– $P_{xy}^a = Pr(x_{t+1} = y | x_t = x, a_t = a)$: the probability of transition from states x to y by action a.

– R: the set of rewards the system can receive from environment, such as:

$$r : \begin{array}{l} (X, X, A) \rightarrow R \\ (x_t, y_t, a_t) \mapsto r_{x_t y_t}^{a_t} \end{array} \qquad [4.5]$$

where r is the reward of environment after applying action a_t during the transition of state $x_t \rightarrow y_t$.

– π: the adopted policy that chooses the action a in state $x : a = \pi(x)$, such as:

$$\pi : \begin{array}{l} X \rightarrow A \\ x \mapsto a \end{array} \qquad [4.6]$$

– Π: the set of policies, such as V^π: the state-value function:

$$\begin{aligned} V^\pi: \quad & (X,\Pi) \rightarrow R \\ & (x,\pi) \mapsto V^\pi(x) \end{aligned} \qquad [4.7]$$

– $V^\pi(x)$: the measure of quality of policy π of state x.

– U: the set of the couples action-state.

– Q^π: the value function of couple action-state, such as:

$$\begin{aligned} Q^\pi: \quad & (U,\Pi) \rightarrow R \\ & (x_0,a,\pi) \mapsto Q^\pi(x_0,a) \end{aligned} \qquad [4.8]$$

The function $Q^\pi(x_0,a)$ measures the quality of the policy π when the system is in the state x_0 and the action a is chosen.

The definition of the functions Q and V (Q-value and V-value) is fundamental because the optimal policy is the one leading to the best quality, which is defined by:

$$\pi^* = \arg\max_{a \in A_x} Q^*(x,a) \qquad [4.9]$$

The optimal policy maximizes the criteria V^π. Therefore, the RL algorithms aim to iteratively improve the function π.

4.2.2. *Value functions*

In this section, we clarify the relationship between the value functions and decision policies.

4.2.2.1. *Rewards*

A reward in a given time r_t is the received signal by the agent during the transition from state x_t to state x_{t+1}, with the realization of action a_t. It is defined by: $r_t = r^a_{xy}$.

The global reward represents the sum of all punctual rewards. It is used to evaluate the system behavior over time. When a trial is finished, the sum is written as:

$$R_{et} = \sum_{k=0}^{T} r_{t+k+1} \qquad [4.10]$$

We define a sum of damped-off rewards by:

$$R_{et} = \sum_{k=0}^{T} \gamma^k r_{t+k+1} \qquad [4.11]$$

where $0 < \gamma < 1$ and T are the end of the test.

4.2.2.2. *Definitions of functions of V-value and Q-value*

The value function V-value represents the expectation of a global sum of punctual rewards using the policy π. The goal of this function is to assess the quality of states under a given policy. It is defined as follows:

$$V^{\pi}(x) = E_{\pi}[R_{et}|x_t = x] = E_{\pi}\left[\sum_{k=0}^{\infty} \gamma^k r_{t+k+1}|x_t = x\right] \qquad [4.12]$$

Similarly, the value function Q-value, $Q^{\pi}(x, a)$, represents the expectation of a global sum when the system realizes the action a under the policy π, such that:

$$Q^{\pi}(x, a) = E_{\pi}\left[\sum_{k=0}^{\infty} \gamma^k r_{t+k+1}|x_t = x, a_t = a\right] \qquad [4.13]$$

The estimated value of these functions is necessary in most RL algorithms. These value functions can be estimated from experiments in [SUT 98] and [LIT 94].

4.2.2.3. *Value function and optimal policy*

A policy $\pi = \pi^*$ is optimal if $V^* = V^{\pi^*}(x) \geq V^{\pi^i}(x)$ with $\forall x \in X$ and for all policies π^i [LIT 94].

The goal of RL is to find an optimal policy π^* that maximizes the expectation of global reward:

$$\pi^* = \arg\max_{\pi} V^{\pi}(x) = \arg\max_{\pi} E_{\pi}\left[\sum_{k=0}^{\infty} \gamma^k r_{t+k+1}|x_t = x\right] \qquad [4.14]$$

The equation [4.12] can be expressed in the form:

$$V^{\pi}(x) = \sum_{y \in X} P_{xy}^{\pi(x)} \left[r_{xy}^{\pi(x)} + \gamma V^{\pi}(y)\right] \qquad [4.15]$$

This equation means that the utility of a state under policy π is an expectation over all available states of the sum of the global reward received by applying $\pi(x)$ and the value of y under the policy π.

This equation also applies to all policies, in particular the optimal policy π^*:

$$V^*(x) = V^{\pi^*} = \sum_{y \in X} P_{xy}^{\pi^*(x)} \left[r_{xy}^{\pi^*(x)} + \gamma V^{\pi^*}(y)\right] \qquad [4.16]$$

By definition, all optimal policies have the same value function V^*. Therefore, the action selected by the optimal policy must maximize the following expression:

$$\sum_{y \in X} P_{xy}^a \left[r_{xy}^a + \gamma V^*(y) \right] \qquad [4.17]$$

which implies:

$$V^*(x) = \max_{a \in A_x} \sum_{y \in X} P_{xy}^a \left[r_{xy}^a + \gamma V^*(y) \right] \qquad [4.18]$$

Finally, we can determine the following optimal policy:

$$\pi^*(x) = \arg\max_{a \in A_x} \sum_{y \in X} P_{xy}^a \left[r_{xy}^a + \gamma V^*(y) \right] \qquad [4.19]$$

These equations require the knowledge of the system model (r_{xy}^a, P_{xy}^a). However, if this model was unknown, the Q-value would solve this problem.

The expression $Q^\pi(x,a)$ can be expressed in the following recurrent form:

$$Q^\pi(x,a) = \sum_{y \in X} P_{xy}^a \left[r_{xy}^a + \gamma V^\pi(y) \right] \qquad [4.20]$$

The function $Q^* = Q^{\pi^*}$ is corresponding to the optimal policy, it does not necessarily choose the action $\pi(x)$ with the policy π.

V^* can be expressed in the following form (optimal equation of Bellman):

$$V^* = \max_{a \in X} \left[Q^*(x,a) \right] \qquad [4.21]$$

and the expression of $\pi^*(x)$ becomes:

$$\pi^*(x) = \arg\max_{a \in X} \left[Q^*(x,a) \right] \qquad [4.22]$$

These last two equations can be evaluated without the knowledge of the system model. Therefore, the problem of RL is learning the Q-values and V-values. In other words, it estimates it in the absence of prior knowledge of the model. One of the efficient methods for solving this problem is the Q-learning we present in the following section.

4.3. Q-learning

Q-learning [WAT 92b] is a learning method that does not require any model. The goal is to find an approximation of the Q-function approaching Q^*. This involves solving the following Bellman optimal equation [MEL 94, BEL 59]:

$$V^* = \max_{a \in A_x} [Q^*(x, a)] \qquad [4.23]$$

The Q-learning algorithm can be summarized in algorithm 4.1. Watkins [WAT 92b] has proved that the values of $Q(x, a)$ reach the optimal values of $Q^*(x, a)$ with the probability of 1.

Algorithm 4.1 Q-learning algorithm

Input: $t = 0$, x: an initial state

repeat

Adjust Q, such as:

$Q(x, a) = Q(x, a) + \alpha \left[r^a_{xy} + \gamma \max_{b \in A_y} Q(y, b) - Q(x, a) \right]$

$x = y$

until $Q \simeq Q^*$

The Q-learning method has obvious advantages. First, it requires no prior knowledge of the system. Second, the perspective of the agent is very local, which is appropriate for the problem of dynamic routing where each router has to decide the routing path without knowing the status of all routers. Furthermore, the implementation requires only a Q-values table comparable to a routing table. This Q-value table is updated after each executed action. However, this algorithm has a drawback that is generally found in all RL methods: this is the guarantee of convergence from Q to Q^*. So, the strong assumptions have to be admitted: the process must be Markovian and stationary, and all couples of state-action (x, a) must be executed an infinite number of times. The last assumption is not crucial in all topics of RL. Indeed, with the problems in robotics where the robot learns to move in a given environment or learning things like learning games (e.g. backgammon and chess), we give a learning period where the agent behavior is irrelevant and where we can afford many explorations. Usually, at the end of this learning period, we continue the exploration leading to better results. In our problem, the environment is not static, but dynamic, and therefore there is no separation between the learning period and operation period. The number of executions of pairs (x, a) must be not only finite, but also enough to ensure the convergence to Q^* without falling into a local minimum. However, it should be small enough so as not to saturate the system. To propose an adaptive routing based on the Q-learning algorithm, the study of a solution to the problem of exploration is developed.

There were already several routing approaches based on Q-learning, called "Q-routing". The difference between these methods and ours is that they used only network parameters as a reward of environment, while we consider user feedback as rewards. The following section presents generally the Q-routing.

4.4. Q-routing

This section discusses the attempt of applying the Q-learning algorithm to adaptive routing protocol, called Q-routing [BOY 94]. The exploitation and exploration framework used in Q-routing is the basis of our proposed routing algorithm.

First, we present the routing information stored at each node as Q-tables. Q-routing is based on Q-learning to learn a representation of a network's state, Q-values, and it uses these values to make control decisions. The goal is to learn an optimal routing policy for the network. The Q-values represent the state s in the optimization problem of routing process. The node x uses its Q-tables Q_x to represent its own view of the network's state. The action a at node x is used to choose neighbor y such that it maximizes the obtained Q-value for a packet destined for node d (destination).

In Q-routing, each node x has a table of Q-values $Q_x(y, d)$, where $d \in V$ is the set of all nodes in the network and $y \in N(x)$ is the set of all neighbors of node x. The $Q_x(y, d)$ is the best Q-value that a packet would obtain to reach the destination d from node x when sent via the neighbor y.

– Exploitation phase: In this phase, the Q-values stored at each node are used for making locally greedy routing decisions. More precisely, when a node x receives a packet $P(s, d)$ destined for node d, node x looks at the vector $Q_x(*, d)$ in its Q-table and selects its neighbor node $\hat{y}$ so that the $Q_x(\hat{y}, d)$ value is maximum[1]. In that way, the node x makes a locally greedy decision by forwarding the packet to the neighbor so that the Q-value is the maximal value.

– Exploration phase: After receiving the packet $P(s, d)$ sent from node x, destined for node d, node y sends back to x its best estimate $Q_y(\hat{z}, d)$ for the destination d: $Q_y(\hat{z}, d) = \max_{z \in N(y)} Q_y(z, d)$. Using the estimate value, the node x updates its $Q_x(y, d)$ value:

$$Q_x(y, d)^{\text{new}} = Q_x(y, d)^{\text{old}} + \alpha(Q_x(y, d)^{\text{est}} - Q_x(y, d)^{\text{old}}) \qquad [4.24]$$

1 In this book, we are working on optimizing the obtained QoE value, so for us the optimized value is the maximal value. Probably, in other approaches where we work on delay optimization, the optimized value is the minimal value.

The goal of Q-routing is to balance the exploitation and exploration phase. In the following section, we survey some related works of Q-routing.

4.5. Related works and motivation

The crucial issue of the success of distributed communication and heterogeneous networks in the next-generation network (NGN) is the routing mechanism. Every routing protocol tries to transport traffic from source to destination by maximizing network performance while minimizing costs. In fact, changes of network environment influence the end user perception [SHA 10a]. For that reason, most existing adaptive routing techniques are designed to be capable of taking into account the dynamics of the network.

The idea of applying RL to routing in networks was first introduced by J. Boyand *et al.* [BOY 94]. The authors described the Q-routing algorithm for packet routing. Each node in the network uses an RL module. They try to keep accurate statistics on which routing decisions lead to minimal delivery time using only local communication. However, this proposal focuses on optimizing only one basis QoS metric: the delivery times. User perception (QoE) is not considered.

The proposals in [MEL 09] and [ESF 08] present two QoS-based routing algorithms based on a multipath routing approach combined with the Q-routing algorithm. The learning algorithm finds K best paths in terms of cumulative link cost and optimizes the average delivery time on these paths. The technique used to estimate the end-to-end delay is based on the Q-learning algorithm to take into account dynamic changes in networks. In the proposal of Hoceini *et al.* [HOC 08], AV-BW Delay Q-routing algorithm uses an inductive approach based on a trial/error paradigm combined with swarm adaptive approaches to optimize three different QoS criteria: static cumulative cost path, dynamic residual bandwidth and end-to-end delay. Based on the K Optimal Q-Routing algorithm (KOQRA), the approach presented here adds a new module to this algorithm dealing with a third QoS criterion which takes into account the end-to-end residual bandwidth.

We can see that the perception and satisfaction of end users are ignored in all of these approaches described above. This poses the problem of choosing the best QoS metric that is often complex. However, QoE comes directly from the user perception and represents the true criterion to optimize. Some other proposals are introduced to solve this problem.

An overlay network for end-to-end QoE management is proposed by B. Vleeschauwer *et al.* [VLE 08]. Routing around failures in the IP network improves the QoE in optimizing the bandwidth usage on the last mile to the client. Components of overlay network are located both in the core and at the edge of the network. However, this proposal does not use any adaptive mechanism.

G. Majd *et al.* [MAJ 09] present a new QoE-aware adaptive mechanism to maximize the overall video quality for the client. Overlay path selection is done dynamically, based on available bandwidth estimation, while the QoE is evaluated using the pseudo-subjective quality assessment (PSQA) tool [RUB 05]. In this model, the video server will decide whether to keep or change video streaming strategies based on end user feedback. However, the adaptive mechanism is quite simple because it is not based on any learning method. Furthermore, using source routing is also a drawback of this approach because this fixed routing is really not adaptive to environment changes.

M. Wijnants *et al.* [WIJ 10] propose a QoE optimization platform that can reduce problems in the delivery path from service provider to users. The proposed architecture supports overlay routing to avoid inconsistent parts of the network core. In addition, it includes proxy components that realize last mile optimization through automatic bandwidth management and the application of processing on multimedia flows. In [LOP 11], the authors propose a wireless mesh networks (WMN) routing protocol based on optimized link state routing (OLSR) to assure both the QoS and the QoE for multimedia applications. They used the dynamic choice of metrics and a fuzzy link cost (FLC) for routing multimedia packets. Definitively, fuzzy system uses two link quality metrics: expected transmission count (ETX) and minimum delay (MD). However, the QoE in these two approaches are considered as a linear combination of QoS criteria. In fact, as mentioned in Chapter 2, QoE is the subjective factor that represents user perception, not a linear combination of QoS parameters.

This book gives a new routing mechanism based on the Q-learning algorithm in order to construct an adaptive, evolutionary and decentralized routing system, which is presented in the following section.

4.6. QQAR routing algorithm

This section presents a new routing algorithm for the routing layer of a CDN architecture, called QQAR protocol. This QoE-aware routing protocol is based on the Q-learning algorithm.

Our idea to take into account end-to-end QoE consists of developing an adaptive mechanism that can retrieve the information from its environment and adapt to initiate actions. The action choice should be executed in response to end users feedback (the QoE feedback). We named this algorithm *QQAR*.

Definitively, the system integrates the QoE measurement in an evolutionary routing system in order to improve the user perception based on the Q-learning algorithm to choose the “best optimal QoE paths” (Figure 4.4). For our method, we have mapped the RL model onto our routing model in the context of a learning routing strategy. So in that way, the routing process is built in order to maintain the best user perception.

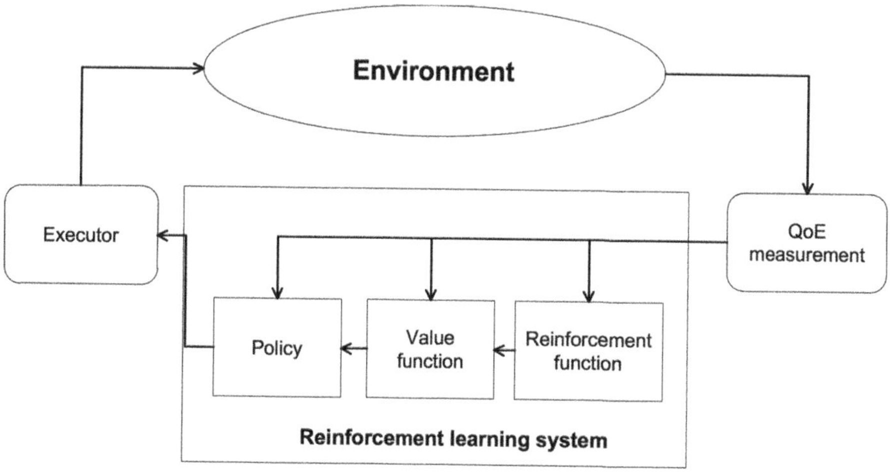

Figure 4.4. *Q-learning-based model*

As mentioned previously, integrating QoE into a system reduces the complexity problem associated with NP-completeness of the QoS model. The objective function is related to user satisfaction, that is the selection of certain parameters, or even one, not several, uncorrelated parameters. We work on the oriented selection gradient on this parameter. In this section, first, we present our formal parametric model based on a continuous time evolving function to prove the convergence of our approach. We then describe in detail the QQAR algorithm. The learning and selection processes explain the way we apply the Q-learning algorithm to QQAR.

4.6.1. *Formal parametric model*

Our system is based on different modules. Each module optimizes a local cost function. To achieve this, we propose a unified formal model to determine the convergence of our approach. Continuous learning in our system involves changing the parameters of the network model using an adaptive update rule as follows:

$$w(n) = w(n-1) - \epsilon(n)F(x(n), w(n-1)) \qquad [4.25]$$

where $w(n)$ denotes the parameters of the whole system at time n, $x(n)$ is the current traffic and ϵ is the modification step. F is either the gradient of a cost function or a heuristic rule. x can be defined by a probability density function $p(x)$ and w is the parameters of the learning system. The concept behind our system is based on the fact that each iteration uses an instance of a real flow instead of a finite training set. The goal of this learning system consists of finding the minimum of a function

$C(w)$, called the expected risk function, and the average update is therefore a gradient algorithm which directly optimizes $C(w)$. For a given state of the system, we can define a local cost function $J(x, w)$ which measures how well our system behaves on x. The goal of learning is often the optimization of some function of the parameters and of the concept to learn which we will call the *global cost function.* It is usually the expectation of the local cost function over the space X of the concept to learn:

$$C(w) = \int_x J(x, w)p(x)dx = E_X\{J(x, w)\} \tag{4.26}$$

Most often we do not know the explicit form of $p(x)$ and therefore $C(w)$ is unknown. Our knowledge of the random process comes from a series of observations x_i of the variable x. We are thus only able to calculate $J(x, w)$ from the observations x_i.

A necessary condition of optimality for the parameters of the system is:

$$\Delta C(w) = E_X\{\Delta_w J(x, w)\} = 0 \tag{4.27}$$

where Δ is the gradient operator. Since we do not know $\Delta C(w)$ but only the result of $\Delta_w J(x, w)$, we cannot use classical optimization techniques to reach the optimum. One solution is to use an adaptive algorithm as follows:

$$w(n) = w(n-1) - \gamma(n)\Delta_w J(x(n), w(n-1)) \tag{4.28}$$

where γ is the gradient step. There is a close similarity between equations [4.25] and [4.28], and when learning in the second stage aims at minimizing a cost function, there is a complete match. This formulation is particularly adequate for describing adaptive algorithms that simultaneously process an observation and learn to perform better. Such adaptive algorithms are very useful in tracking a phenomenon that evolves over time, as in user perception.

4.6.2. *QQAR algorithm*

Our approach is based on the Q-learning algorithm. In this section, we summarize some general idea of RL and Q-learning.

Q-learning [WAT 92a] is a technique for solving specific problems by using the RL approach [SUT 98]. Even though the RL has been around for a long time, this concept has been applied/used recently in many works in many disciplines, such as game theory, simulation-based optimization, control theory and genetic algorithms.

Here we cite here some recent RL advances: RL-Glue [TAN 09], PyBrain [SCH 10], Teachingbox [ERT 09], etc.

Here we explain how an RL framework works. In an RL framework, an agent can learn control policies based on experience and rewards. In fact, MDP is the underlying concept of RL.

An MDP model is represented by a tuple $(\mathcal{S}, \mathcal{A}, \mathcal{P}, \mathcal{R})$, where $\mathcal{S}$ is the set of states, $\mathcal{A}$ is the set of actions and $\mathcal{P}(s'|s, a)$ is the transition model that describes the probability of entering state s' after executing action a at state s. $\mathcal{R}(s, a, s')$ is the reward obtained when the agent executes a at state s and enters s'. The quality of action a at state s is evaluated by Q-value $Q(s, a)$.

Solving an MDP is equivalent to finding an optimal policy $\pi : \mathcal{S} \rightarrow \mathcal{A}$ that maps states to actions such that the cumulative reward is maximized. In other words, we have to find the following Q-fixed points (Bellman equation):

$$Q^*(s,a) = r(s,a) + \gamma \sum_{s'} P(s'|s,a) max_{a'} Q^*(s',a') \qquad [4.29]$$

where the first term on the right side represents the expected immediate reward of executing action a at s; the second term represents the maximum expected future reward and γ represents the discount factor.

One of the most important breakthroughs in RL is an off-policy temporal difference (TD) control algorithm known as Q-learning, which directly approximates the optimal action-value function (equation [4.29]), independent of the policy being followed. After executing action a, the agent receives an immediate reward r from the environment. By using this reward and the long-term reward it then updates the Q-values influencing future action selection. One-step Q-learning is defined as:

$$Q(s,a) = (1-\alpha)Q(s,a) + \alpha[r + \gamma \max_{a'} Q(s',a')] \qquad [4.30]$$

where α is the learning rate, which models the rate of updating Q-values.

In order to apply the Q-learning algorithm to our routing system, we have mapped the RL model onto our routing model in the context of learning a routing strategy. We consider each router in the system as a state. The states are arranged along the routing path from the client to the chosen server in a CDN. Furthermore, we consider each link emerging from a router as an action to choose. The system's routing mechanism refers to the policy π (Table 4.1). Regarding QoE measurement, we used a method that evaluates the QoE of all nodes.

RL-model	Routing model
State	Router
Actions	Links
Policy π	Routing policy

Table 4.1. *Mapping RL-model onto routing model*

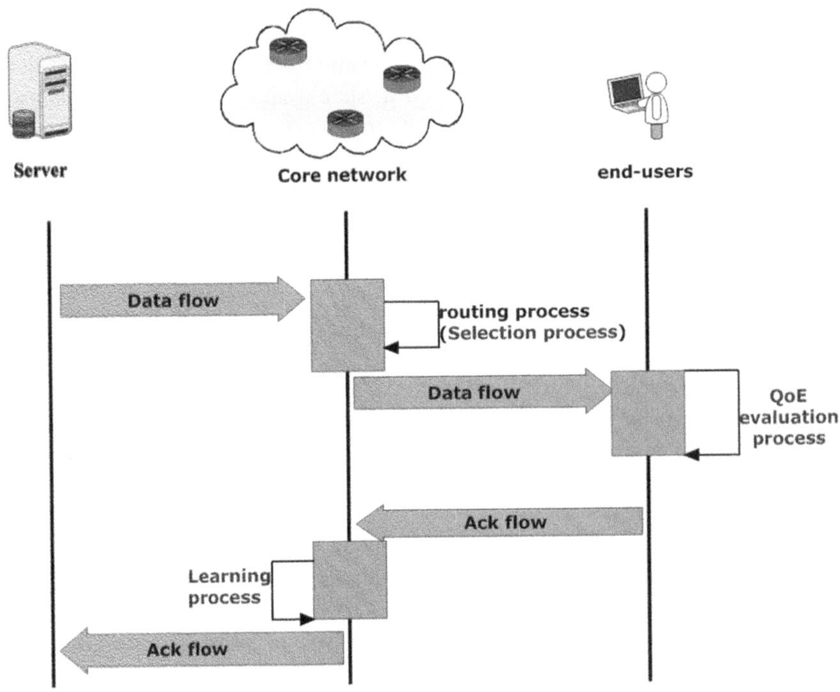

Figure 4.5. *QQAR protocol*

The routing mechanism is illustrated in Figure 4.5

– *First step – Data packet flow*: the server chosen by meta-routing layer sends a data packet to the end user. After receiving this data packet, each node in the routing path evaluates QoE and forwards the packet to the next path in using a selection process (presented in section 4.6.5).

– *Second step – On the user side*: after data reach end users, the QoE evaluation process is realized to send back a QoE feedback as an ACK message to the routing path.

– *Third step – ACK message flow*: once a node receives an ACK message, it updates the Q-value of the link that this ACK message just passed through. The update process is introduced below (section 4.6.3). It then attaches the QoE measurement result to the ACK message and forwards it to the previous neighbor.

Figure 4.6 describes the QQAR algorithm.

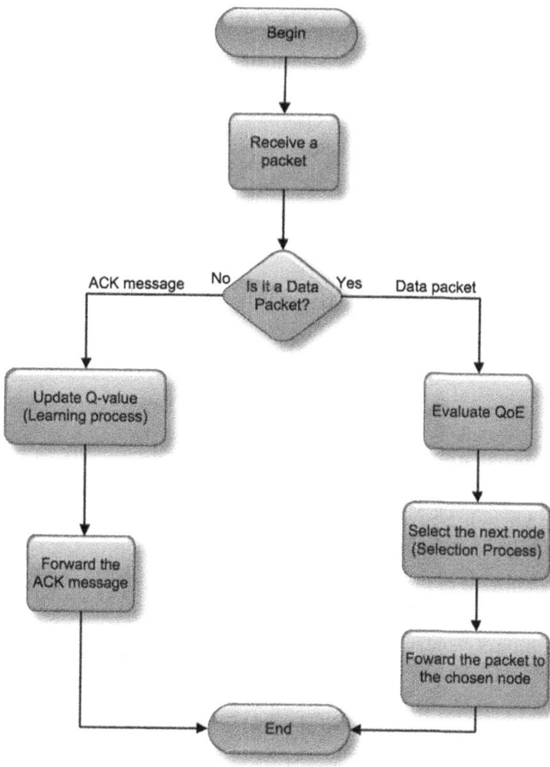

Figure 4.6. *QQAR diagram*

In the following two sections, we clarify the two main processes of QQAR: the learning process and the selection process.

4.6.3. *Learning process*

As the analysis of Q-learning algorithm presented previously, we present a new method for network exploration. This technique relies on forward exploration [BOY 94] to which we added a probabilistic mechanism exploration. This mechanism allows us, first, to explore from time to time optimal paths and, second, to avoid further congestion of the network as data packets are also used for exploration.

In our model, each router has a routing table that indicates the Q-values of links emerging from this router. For example, in Figure 4.7, node y has a routing table containing Q values: $Q_{yz_1}, Q_{yz_2}, Q_{yz_3} ... Q_{yz_n}$ corresponding to n links from y to z_i with $i = 1, \ldots, n$. Based on this routing table, the optimal routing path can be trivially constructed by a sequence of table look-up operations. Thus, the task of learning optimal routing strategy is equivalent to finding the Q-fixed points (equation [4.29]).

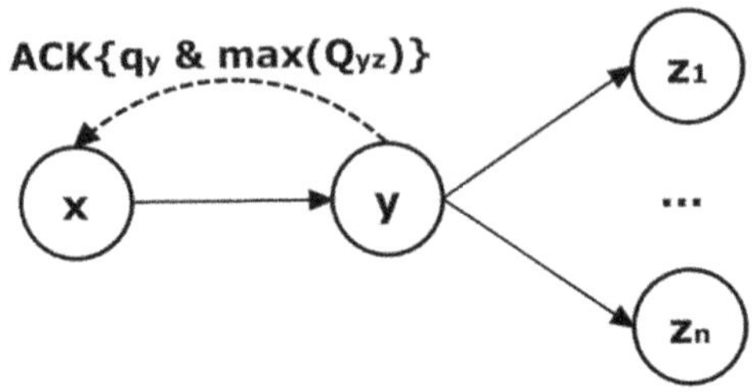

Figure 4.7. *Learning process*

As mentioned above, after receiving a data packet, the destination node (end user) evaluates the QoE. Then, it sends back the feedback (QoE value) as ACK message to the routing path. Each router x in the routing path receives this message, including the QoE evaluation result (the MOS score[2]) of the previous router (q_y) and the maximum value of Q-values ($\max Q_{yz}$) in the routing table of node y (Figure 4.7).

Based on the native Q-learning algorithm (equation [4.30]), the update function is expressed as:

$$Q_{\text{new}} = Q_{\text{old}} + \alpha \left[Q_{\text{new estimation}} - Q_{\text{old}}\right] \qquad [4.31]$$

where Q_{new} and Q_{old} are the new and old Q-values, respectively. The $Q_{\text{new estimation}}$ is the new estimation of Q-values from nodes x to the destination, depending on the following two factors:

1) The estimation of the link xy, expressed by the difference between the Q-values of node x and y. We call that the *immediate reward*. The reason of choosing this estimation is explained below.

2 Mean opinion score (MOS) gives a numerical indication of the perceived quality of the media received after being transmitted. The MOS is expressed in numbers, from 1 to 5, 1 being the worst and 5 the best. The MOS is generated by averaging the results of a set of standard, subjective tests where a number of users rate the service quality.

2) The optimal estimation from node y to the destination, expressed by the maximal value among the Q-values in the Q-table of node y. We call that the *future reward.*

So, we have the Q-value of the new estimation expressed in equation [4.32]:

$$Q_{\text{new estimation}} = \underbrace{\beta(q_y - q_x)}_{\text{immediate reward}} + \underbrace{\gamma \max_i Q_{yz_i}}_{\text{future reward}} \qquad [4.32]$$

where β and γ are two discount factors balancing the values between *future reward* and *immediate reward.*

From two equations [4.31] and [4.32], router x updates the Q-value of the link xy by using the update function defined in equation [4.33]:

$$\underbrace{Q_{xy}}_{\text{new value}} = \underbrace{Q_{xy}}_{\text{old value}} + \alpha \Big[\underbrace{\beta(q_y - q_x) + \gamma \max_i Q_{yz_i}}_{\text{new estimation}} - \underbrace{Q_{xy}}_{\text{old value}} \Big] \qquad [4.33]$$

where Q_{xy} and Q_{yz_i} are Q-values of links xy and yz_i. q_x and q_y are QoE value evaluated (MOS score) by using the QoE measurement method at node x and y. α is the learning rate, which models the rate updating Q-value. Actually, it is worthwhile noting here the close similarity between equation [4.33] and [4.28] given in the global cost function. Parameters α and Q represent the gradient step γ and the function J in the formal model respectively (equation [4.28]). In fact, QQAR becomes reactive with the update function in equation [4.33]. The router will reduce the Q-value of a link when its quality is degraded. This reduction helps QQAR to actively avoid bad quality links.

Algorithm 4.2 Learning algorithm

Input: $[Q_0, Q_1, Q_2, ..., Q_m]$: Q-values in Q-table.
m: The number of neighbor nodes.
N: the number of node of the routing path.
for $i = 0$ to N **do**
  Receive the ACK message;
  Update the Q-value;
  Find the maximal Q-value and attach it to the ACK message;
  Forward the ACK message;
end for

According to the time complexity of the learning algorithm (algorithm 4.2), the learning process is executed N times with N the number of node throughout the routing path. In addition, the number of operations for finding maximal Q-value in the Q-table is m, which is the number of neighbors of this node. As shown in algorithm 4.2, the number of elementary operations performed by the learning algorithm for a loop is $(m + 3)$. In fact, there are totally N loops. So, the asymptotic time complexity of the learning algorithm is $O(N \times m)$.

4.6.4. *Simple use case-based example of QQAR*

We now show some samples of execution of QQAR. We consider a simple network topology as shown in Figure 4.8. This network consist of 11 routers ($R1$, $R2$, ... , $R11$), one source node, one destination node (the chosen server in a CDN) and 14 connection links. With this sample network, we have three paths from source to destination: $R1$-$R2$-$R3$-$R4$-$R5$-$R6$, $R1$-$R7$-$R8$-$R9$ and $R1$-$R11$. The path $R1$-$R7$-$R8$-$R10$ never reaches the destination. All routers used the QQAR algorithm. We fixed the three parameters $\alpha = \beta = \gamma = 0.9$. In order to show the execution of the exploration phase easily, we have chosen this high value for these parameters to make the learning change rapidly. Note that the numbers appearing on the links in these figures are their Q-values.

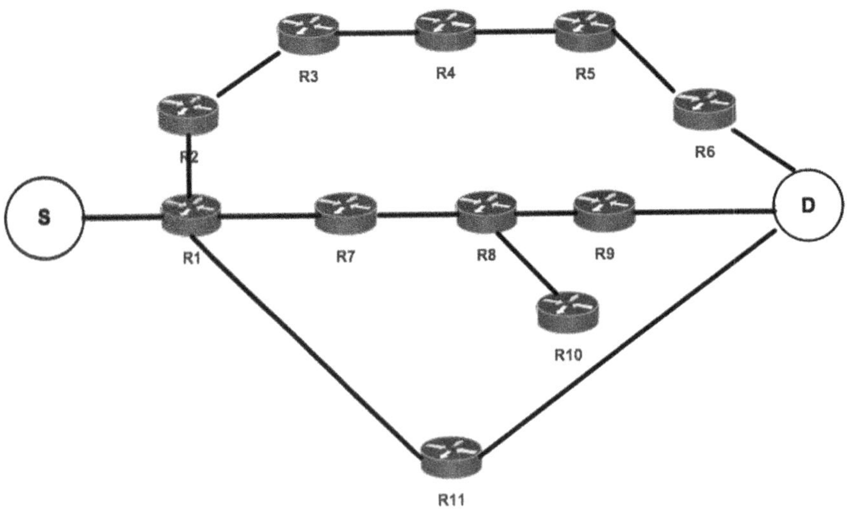

Figure 4.8. *Sample of execution for QQAR*

In the first phase (Figure 4.9), assume that the path $R1$-$R2$-$R3$-$R4$-$R5$-$R6$ is chosen and the destination gives an MOS score of 4. We can see that after the update process, all the links of the routing path receive a high Q-value.

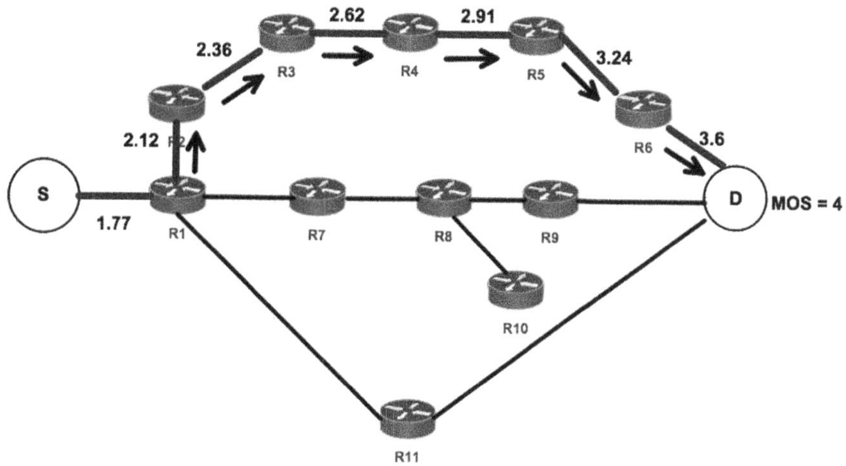

Figure 4.9. *Phase 1: with MOS = 4, the Q-values are updated*

In the second phase (Figure 4.10), the path *R*1-*R*7-*R*8-*R*9 is chosen. The destination has given an MOS score of 3.

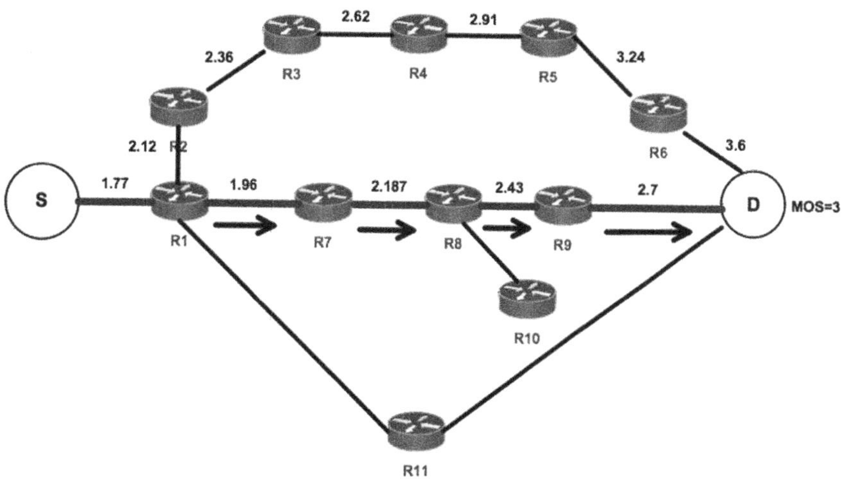

Figure 4.10. *Phase 2: with MOS = 3, the Q-values are updated*

The path *R*1-*R*11 has been chosen in third phase (Figure 4.11). The destination has given a MOS score of 1. The Q-values in this routing path are updated by using equation [4.33]. This phase represents the case where a short routing path in terms of hop-count will have low QoE value if the end user is not satisfied with the received quality.

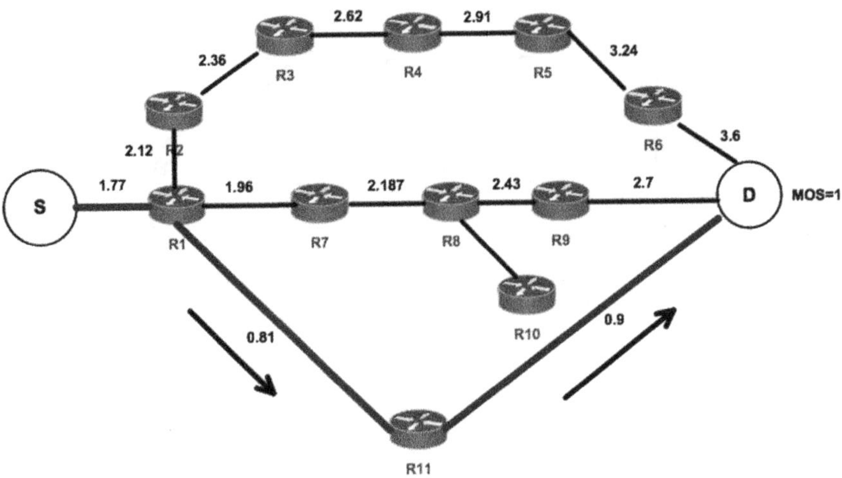

Figure 4.11. *Phase 3: with MOS = 1, the Q-values are updated*

In the fourth phase, the path *R*1-*R*7-*R*8-*R*10 is chosen (Figure 4.12). The packet cannot reach the destination, so the QoE value obtained is 0. We can see that our update function of equation [4.33] does not influence the Q-value in the paths *R*1-*R*7 and *R*7-*R*8. Therefore, the router R8 will never choose the path *R*8-*R*10 to forward the packet.

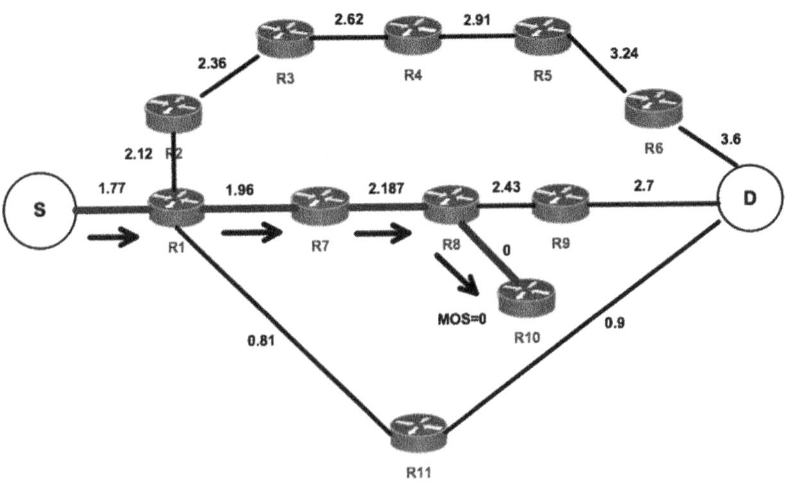

Figure 4.12. *Phase 4: with MOS = 0, the Q-values are updated*

In the fifth phase, the path $R1$-$R2$-$R3$-$R4$-$R5$-$R6$ is chosen for the second time (Figure 4.13). We assume that the path quality is degraded compared with the first time. So, the MOS value given is just 2. The update process has decreased all Q-values of this path.

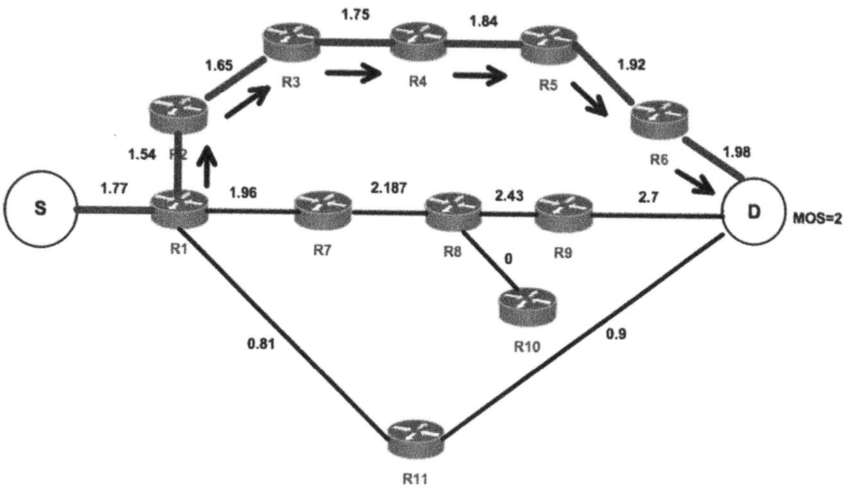

Figure 4.13. *Phase 5: with MOS = 2, the Q-values are updated*

The samples of algorithm execution above show some characteristics of exploration process of QQAR:

– If path $P1$ obtains a better QoE value (MOS score) than path $P2$, all the links of path $P1$ obtain better Q-values than $P2$ (phase 1 and 2).

– A path has low Q-values if it obtains a poor quality at end user even if it is close to the destination in terms of hop-count (phase 3).

– If a path cannot reach the destination, this path will not be chosen in the future. (phase 4).

– The decreasing of the MOS score at the end user will decrease all Q-values of the corresponding routing path (phase 5).

In order to explain why the new estimation in equations [4.32] and [4.33] depends on the difference of Q-values between two adjacent nodes, $(q_y - q_x)$, we give another sample network in Figures 4.14 and 4.15. We keep the value of 0.9 for three parameters α, β and γ. This network includes nine routers, one destination and 16 connection links. The router $R4$ is connected to all other routers. We make router $R4$ connect to all other routers in order to analyze the Q-routing issue that is solved by our proposed difference in formulation [4.33].

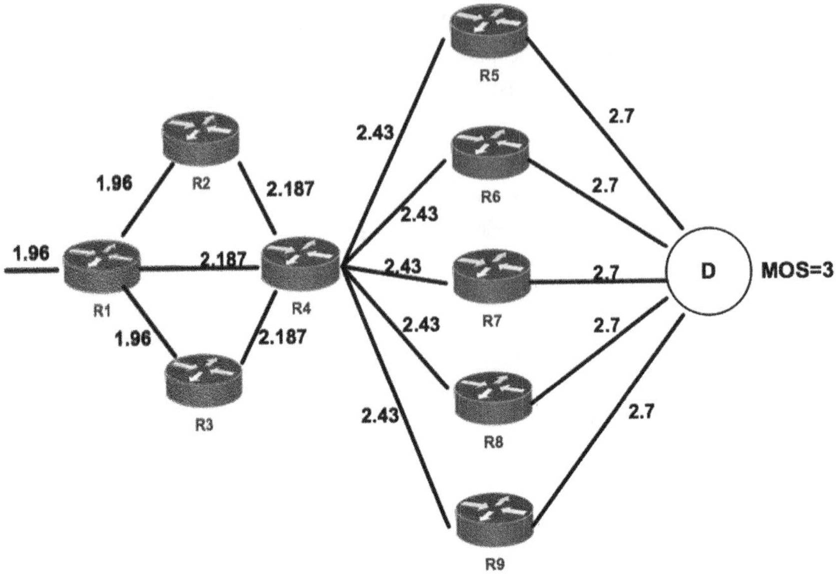

Figure 4.14. *Network before the failure*

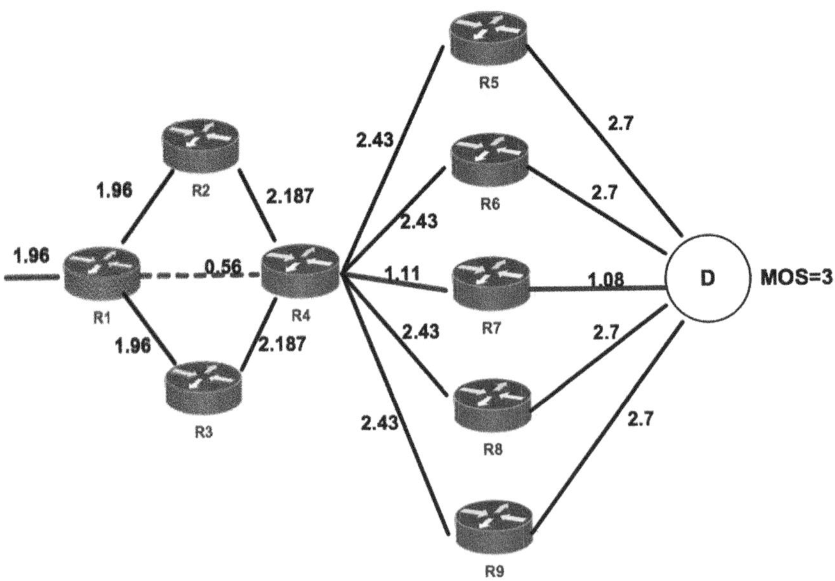

Figure 4.15. *QQAR detects exactly the downed link*

In fact, due to the difference in the update function, the QQAR is able to detect exactly where the failure comes from. Figure 4.14 describes network state before the failure. With the MOS score of 3 obtained at the destination, all links have their own Q-values assigned by QQAR algorithm. When the failure occurs in the link $R1$-$R4$ (Figure 4.15) and the routing path is $R1$-$R4$-$R7$, the MOS score decreases from 3 to 1. The router $R7$ and $R4$ update the Q-values of links $R7$-D and $R4$-$R7$, which are equal to 1.08 and 1.11, respectively. We now analyze the update process of router $R1$. Without the difference of q_{R4} and q_{R1}, the update function becomes:

$$\underbrace{Q_{\text{R1-R4}}}_{\text{new value}} = \underbrace{Q_{\text{R1-R4}}}_{\text{old value}} + \alpha \left[\underbrace{\gamma \max_i Q_{\text{R4-Ri}}}_{\text{new estimation}} - \underbrace{Q_{\text{R1-R4}}}_{\text{old value}} \right] \qquad [4.34]$$

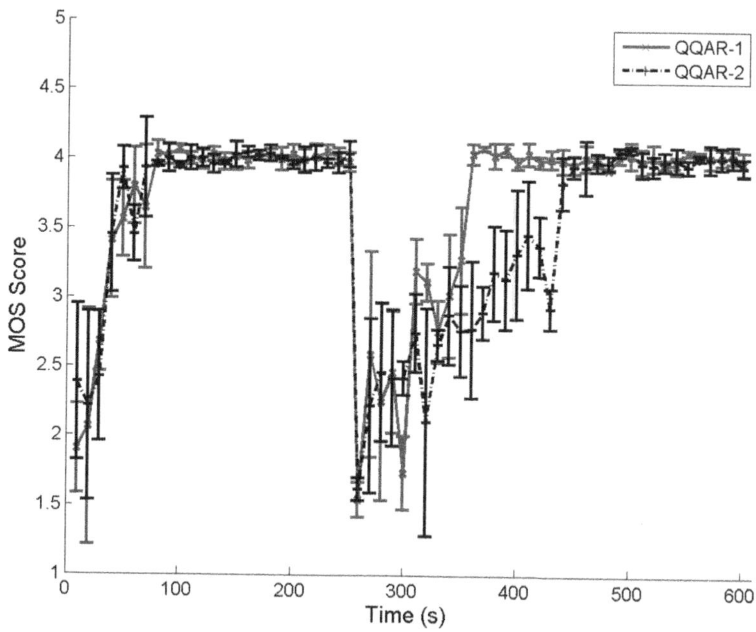

Figure 4.16. *Average MOS score of QQAR-1 and QQAR-2*

Since the highest value of Q-values of links emerging from $R4$, $\max_i Q_{\text{R4-Ri}}$, is still 2.43, the Q-value of link $R1$-R4 is still 2.187. In reality, the problem of quality degradation comes from the link $R1$-R4. In order to solve this problem, we added the difference of Q-values between two adjacent routers (equation [4.33]). Adding this difference helps us to find the link that causes the quality degradation problem. In fact, this difference makes the Q-value of link $R1$-R4 decrease from 2.187 to 0.56 (Figure 4.15). With this new low Q-value, the future routing paths of QQAR will avoid this low-quality link.

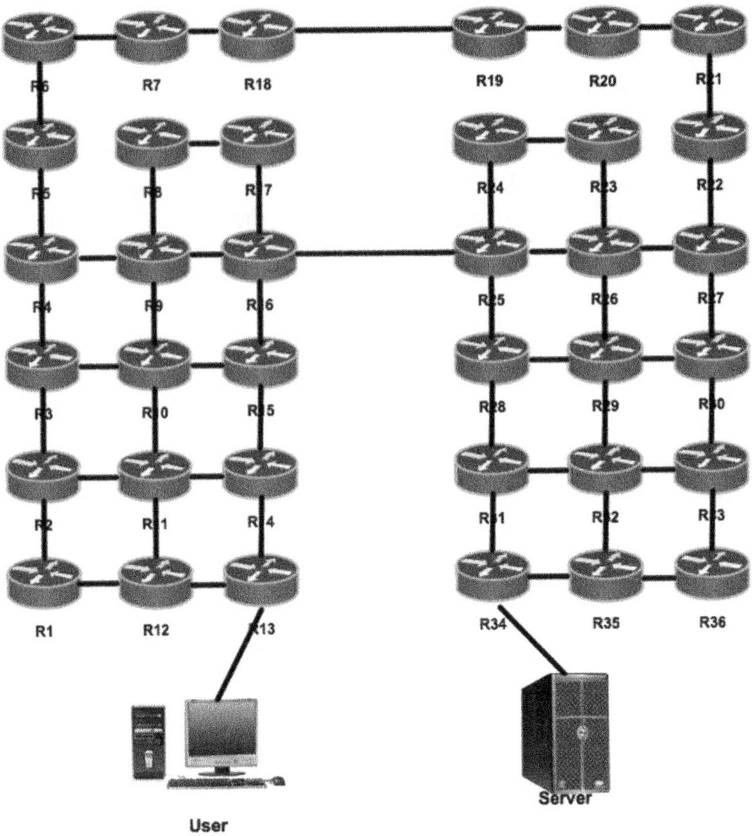

Figure 4.17. *Irregular* 6×6 *grid network*

In order to prove this advantage, we have tested two versions of the algorithm QQAR: *QQAR*-1 using formulation [4.33] as an update function and *QQAR*-2 using formulation [4.34] to update their Q-values. The only difference between them is the difference $(q_y - q_x)$. We used the irregular network topology 6×6 grid (Figure 4.17). The test lasts 10 min. After 250 s, we make link *R*16-*R*25 break down to see how these two versions of QQAR react to this problem.

As shown in Figure 4.16, after approximately 80 s of the initial period where two algorithms learn the network topology, they obtain an average MOS score of 4.0. At the 250th min, the MOS score obtained by two algorithms decreases to 1.5 since the link *R*16-*R*25 is broken down. Subsequently, while *QQAR*-2 takes 220 s to regain the MOS value of 4.5, the *QQAR*-1 takes just about 100 s. Therefore, in this example, the capacity of QQAR with the difference of Q-value between two adjacent nodes for reacting to the environment changes is better than the version without this difference.

So, with this efficient learning process, all nodes can use their updated Q-values in the Q-table to select the next node for routing the packet received. The selection process is presented in the following section.

4.6.5. *Selection process*

With regard to the selection process in each node after receiving a data packet, we have to consider the trade-off between *exploration* and *exploitation*. It cannot always *exploit* the link that has the maximum Q-value because each link must be tried many times to reliably estimate its expected reward. This trade-off is one of the challenges that arises in RL and not in other kinds of learning. To obtain good rewards, an RL agent (router) must prefer actions that it has tried in the past and found to be effective in producing reward. But to discover such actions, it tries actions that have not selected before. The agent has to *exploit* what it already knows, but it also has to *explore* in order to make better action selection in the future. There are some mathematical issues to balance exploration and exploitation. We need a method that varies the action probabilities as a graded function of estimated value. The greedy action (i.e. the action that gave the best quality feedback in the past) still has the highest selection probability. All the other actions are ranked and weighted according to their value estimates. Therefore, we have chosen the *softmax method* based on the Boltzmann distribution [SUT 98].

With these softmax action selection rules, after receiving a packet, node x chooses a neighbor y_k among its n neighbors y_i $(i = 1, \ldots, n)$ with probability expressed by equation [4.35] to forward the packet:

$$p_{xy_k} = \frac{e^{\frac{Q_{xy_k}}{\tau}}}{\sum_{i=1}^{n} e^{\frac{Q_{xy_i}}{\tau}}} \quad (1 \leq k \leq n) \qquad [4.35]$$

where Q_{xy_i} is the Q-value of link xy_i and τ is the temperature parameter of Boltzmann distribution. High temperature makes the link selection be all equi-probable. On the contrary, low temperature generates a greater difference in selection probability for links that differ in their Q-values. In other words, the more we reduce the temperature τ, the more we exploit the system. Initially, we begin by exploring the system with the initial value of the temperature τ. Then, we increasingly exploit the system by decreasing this temperature value. In that way, we reduce τ after each forwarding time as shown in equation [4.36]:

$$\tau = \phi \times \tau \quad (0 < \phi < 1) \qquad [4.36]$$

where ϕ is the weight parameter.

More precisely, the selection process can be summarized in algorithm 4.3:

Algorithm 4.3 Selection algorithm

Input: $[Q_0, Q_1, Q_2, ..., Q_n]$: Q-values in Q-table.
repeat
Receive the data packet;
Look at the Q-table to determine the Q-value for each link;
Calculate the probabilities with respect to Q-values (equation [4.35]);
Select link with respect to the probabilities calculated;
Decrease the temperature value (equation [4.36]);
until Packet reaches the destination

We now analyze the algorithm complexity of the selection algorithm (algorithm 4.3). There are N loops in total, where N is the number of node in the routing path. Calculating the probabilities for each link has a time complexity of $O(m)$, where m is the number of neighbors. As $N \leq \mathbb{N}_{\text{TTL}}$, where $\mathbb{N}_{\text{TTL}}$ is *time to live* we fixed before (i.e. the number of hops the packet is allowed to pass before it dies), the worst-case time complexity for the selection algorithm is $O(\mathbb{N}_{\text{TTL}} \times m)$.

4.7. Experimental results

In order to validate our studies, this section shows the experimental results of our two proposals: the routing algorithm QQAR and the new server selection method.

4.7.1. *Simulation setup*

We also implement a network simulation model to validate the proposals. First, we present the simulation setup including setting parameters, simulator and simulation scenario.

Table 4.2 lists the setting parameters.

4.7.1.1. *Simulator*

There are many network simulators (NSs) allowing the implementation and performance evaluation of network systems and protocols. They are either prototypes from academic research or marketed products. In this section, we present three common simulators: NS, MaRS and the optimized network engineering tool (OPNET). Subsequently, we present in detail the OPNET and explain why we have chosen these simulators to simulate our proposals.

4.7.1.1.1. NS

NS [NET 89] is distributed by the University of Berkeley (USA). This is a discrete event simulator designed for research in the field of networks. It provides substantial support for simulations using TCP and another routing protocols. The simulator is written in *C++* and uses *OTcl* as the language interface and configuration control.

OS	Windows XP 32bit
CPU	Intel Core 2 6700 2.66 GHz
RAM	2.0 GB
OPNET Version	Opnet Modeler v14.0
Topologies	6 × 6 grid 3AreasNet NTT
6 × 6 **grid network**	Router number: 36 User/iBox number: 3 Server number: 2 Link number: 50
3AreasNet network	Router number: 35 User/iBox number: 10 Server number: 5 Link number: 63
NTT network	Router number: 57 User/iBox number: 10 Server number: 5 Link number: 162
Simulation time	45 h
Confidence interval	95%
Arrival request distribution	Markov-Modulated Poisson Process (MMPP)

Table 4.2. *Simulation setup*

The advantage of NS is that it offers many extensions to improve its basic characteristics. The user specifies the network topology by placing the list of nodes and links in a topology file. Traffic generators can be attached to any node to describe its behavior (e.g. FTP or Telnet client). There are different ways to collect information generated during a simulation. Data can be displayed directly during the simulation or be saved to a file for processing and analyzing.

The two monitoring modes in NS are, first, the registration of each packet as it is transmitted, received and rejected and, second, the monitoring of specific parameters, such as the date of arrival of packets and size. Data generated during the simulation are stored in files and can be treated. The *Network Animator* is a tool for analyzing these files. Its graphical interface displays the configured topology and packet exchanges.

The NS is an open source simulator (its source code is available in [NET 89]). It is widely used for research in the field of networks, and recognized as a tool for experimenting with new ideas and new protocols.

4.7.1.1.2. MaRS

The Maryland routing simulator (MaRS) [MAR 91] is a discrete event simulator that provides a flexible platform for the evaluation and comparison of routing algorithms. It is implemented in language *C* under UNIX, with two GUIs: *Xlib* and *Motif.*

MaRS allows the user to define a network configuration consisting of a physical network, routing algorithms and traffic generators. The user can control the simulation, record the value of certain parameters and save, load and modify network configurations. MaRS consists of two parts: (1) a simulation engine, which controls the event list and the user interface; (2) a collection of components to model the network configuration and to manipulate certain features of the simulation. The configuration of the simulation (network algorithms, connections and traffic) can be done via configuration files or via a graphical *X-Windows.*

4.7.1.1.3. OPNET

The OPNET project was launched in 1987. Currently, this software is marketed by MIL3 and is the first commercial tool for simulation of communication networks.

This is a simulation tool based on discrete events, written in *C*. It has a graphical interface used in various modes, which facilitates the development of new models and simulation programs. Many models of protocols are provided with the standard distribution of OPNET, which is a definite advantage over other simulators. In fact, the other simulators often offer only an implementation of *TCP/IP* and some classical routing protocols. OPNET is a very flexible tool, which simulates a network at any level of granularity. It is used extensively in many development projects.

The advantages of this simulator tool motivate us to choose it in order to simulate and validate our proposals in this book. The following section presents, in detail, simulation settings using OPNET simulator.

4.7.1.2. *Simulation scenarios*

The Internet today has a dynamic environment with lots of unexpected changes that cause different network issues. The *load balance* is an important factor that influences the network performance. There are different load balancing methods that have been introduced to avoid this issue. In fact, overloaded routers or servers can break down the entire network system rapidly. In addition, *unexpected incidents* also affect the network service quality. A failed router or link can change the network topology. This change really decreases the network quality and requires an effective fault-tolerance mechanism of the network operator. Therefore, we simulate the following two scenarios:

1) *High-load scenario*: We have generated traffic that stresses the network. The load level represents the rate of the number of overloaded links and number of total links:

$$\text{level} = \frac{n_s}{N} \times 100\% \qquad [4.37]$$

where n_s is the number of loaded links and N is the total number of links in the system. Therefore, the more we increase the *level*, the more network system is loaded. We test it by using seven levels: 10%, 20%, 30%, 40%, 50%, 60% and 70%.

Furthermore, we increase the bandwidth of all links in the network. As shown in the equation [4.38], the new bandwidth is assigned the new value:

$$\text{BW}^i_{\text{new}} = \text{level} \times \text{BW}^i_{\text{old}} \qquad [4.38]$$

where the BW^i_{new} and BW^i_{old} are respectively the new and old values of bandwidth of link i. The level is varied from 10% to 70% as mentioned above.

2) *Incident scenario*: During the simulation time, we make routers or servers randomly break down. The time duration of this broken phase is long enough to test the fault tolerance capacity of the approaches. This mechanism can evaluate the capacity of detecting and reacting to broken issues, which represents the dynamic of the approaches to the environment changes. Given a router X, Figure 4.18 illustrates the time line for incident scenarios for this router. The broken time lasts from t_x to t_y. In this broken period, the router X neither receives nor forwards packets. The t_x and $T_{\text{broken time}}$ are chosen randomly so that the $T_{\text{broken time}}$ is long enough to test the reacting method of the approach.

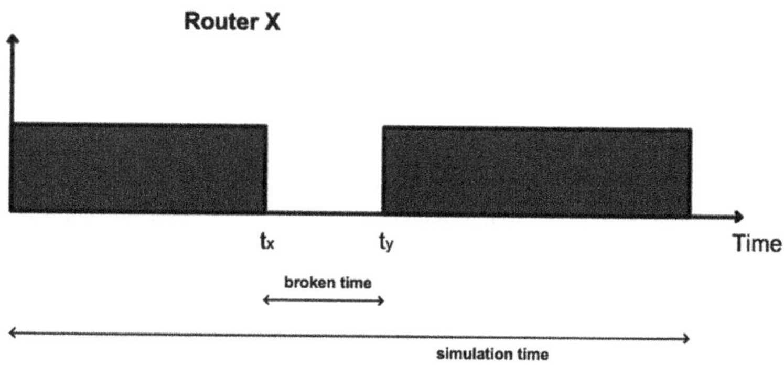

Figure 4.18. *Incident time line for router X*

4.7.1.3. *Simulation metrics*

In this book, we used the following metrics to evaluate the performance of the approaches:

– Average MOS score: This metric gives the overall MOS score of the entire network system. It is equal to the average obtained MOS score on the client side. Equation [4.39] is used to calculate the average MOS score:

$$\text{Average MOS score} = \frac{\sum_{i=1}^{n} m_i}{n} \qquad [4.39]$$

where n is the number of evaluated MOS scores and m_i is the evaluated value at the ith time.

– Packet delivery ratio (PDR): This metric gives the ratio of the data packets successfully received at the destination node and the total number of packets sent at the source node. We used equation [4.40] to calculate the PDR:

$$\text{PDR} = \frac{D_R}{D_S} \times 100 \qquad [4.40]$$

where D_R is the data packets received by the destination and D_S is the data packets sent by the source.

– Normalized Routing Overhead (NRO): This metric is the number of control packets (e.g. error, request and acknowledge) transmitted per data packet delivered to the destination. Equation [4.41] is used to calculate the NRO:

$$\text{NRO} = \frac{(P_S + P_F)}{D_R} \qquad [4.41]$$

where P_S is the control packets sent by all nodes and P_F is the control packets forwarded by all nodes. D_R is the data packets received by the destination.

4.7.1.4. *Dynamic network system*

One of the most important capacities of a routing algorithm is adapting to changes in crucial system parameters during network operation. In addition, the Internet today always changes dynamically. Therefore, we tested our approaches on networks whose topology, end user requests distribution and traffic load levels were dynamically changing.

– *Topology*: some routers in the network will break down for a period of time to change the network topology. This is a challenge for the algorithms to react quickly to such changes and continue working efficiently.

– *End user request distribution*: The request distribution is based on Markov-modulated Poisson Process (MMPP) distribution, which is presented below (section 4.7.1.5.3).

– *Traffic load*: We use an automatic method to dynamically change the load of links in the network by varying their bandwidth. The way to change bandwidth was discussed in section 4.7.1.2. The effect of load levels is shown more clearly in experimental results.

4.7.1.5. *Simulation settings*

For performance validation of the proposals, we have chosen the simulator OPNET for several reasons. It allows us to visualize the physical topology of a local, metropolitan, remote or embedded network. The protocol-specified language is based on the formal description using finite state automata. It is also easier to understand the problems of transit times and network topology.

OPNET has three nested hierarchical levels: the *network domain*, the *domain node* and the *process domain*. In the following sections, we present our simulation settings of these levels.

4.7.1.5.1. Network domain

This is the highest level of the hierarchy of OPNET. It defines the network topology by deploying hosts, links and active equipment such as switches or routers. Defined by its model, each communication entity (called node) is configurable. The graphical user interface associated with the network domain takes into account the spatial deployment of the network. According to the network, we worked on the three following topologies: 6×6GridNet, 3AreasNet and NTT. We have chosen a testing methodology as follows.

– 6×6*GridNet* (Figure 4.19b): First, we tested our routing algorithm on a simple network, the 6×6GridNet, to verify its functionality. This irregular network was first used by Boyan and Littman [BOY 94]. Then, it was widely used and accepted in routing protocols studies. The network consists of two connected areas with a bottleneck of traffic on two bridging links. After obtaining sufficient results for the first topology, we continued testing our algorithm by using the second topology, the 3AreasNet.

– *3AreasNet* (Figure 4.19a): For the second testing step, we ourselves have designed a simulated irregular network, called 3AreasNet, with three separated areas including 35 routers. Each area is connected to each other by a unique link and all links are similar. This constitutes our second beta test bed, which is used to test the functionality of algorithms. The 3AreasNet topology is based on Waxman topology [WAX 93]. This method distributed the node uniformly in the network, and the edges are added considering the probability defined by equation [4.42] that depends on the distances between nodes.

$$P = \alpha \times a \exp\left(-\frac{d}{L} \times \beta\right) \qquad [4.42]$$

where d represents the distance between two nodes, L is the maximum distance in the network, $\alpha > 0$ and $\beta <= 1$. The parameter α increases or decreases the probability of edges between two nodes and β controls the ratio of long/short edges.

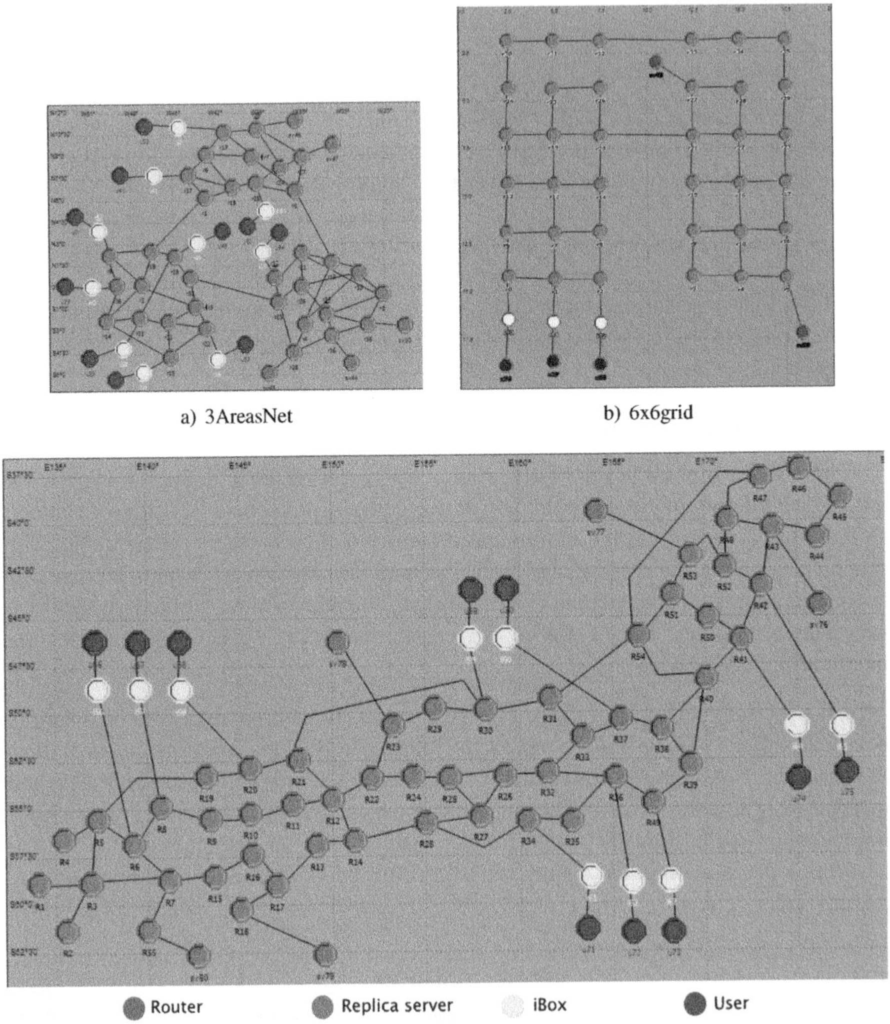

a) 3AreasNet

b) 6x6grid

c) NTTnet

Figure 4.19. *Network topologies. For a color version of the figure, see www.iste.co.uk/mellouk/qoe*

– *NTT* [DIC 05]: For the third step, we worked on the NTT network that is modeled on the exact characteristics of a real-world network, the former Nippon Telephone and Telegraph (NTT) company fiber-optic corporate backbone. NTTnet is a 57 node, 162 bi-directional link network (Figure 4.19c). Unlike the 3AreasNet and 6 × 6GridNet topologies that are used to test the functionality of algorithms, this topology is used to validate the proposed approaches.

In each network, we used 10 end users (red color), 10 iBoxes (yellow color) and five replica servers (green color). According to the server placement, we used a server placement method proposed in [QIU 01]. This method is based on a greedy algorithm [KRI 00]. For this method, we have to choose five servers in potential sites. The site having the lowest cost (hop-count to end users) is chosen. Then, the iteration continues until five servers have been chosen. According to the cache organization, we assume that all requested data is stored in each server.

4.7.1.5.2. Node domain

Node domain is used to define the formation of nodes (e.g. routers, workstations and hubs). The model is defined by using blocks called modules. Some modules are not programmable: it is primarily *transmitters* and *receivers*, whose unique function is to interface the node and links. On the other hand, other modules are programmable such as *processors* and *queues*.

Processors are modules that perform a specific task node. They symbolize the different parts of the operating system of a machine, and the different layers mainly implemented in the node network (e.g. Ethernet and IP). In addition, these modules can communicate with each other via packet streams, which allow a packet to transit from one layer to another within the same machine. This kind of organization provides a clear vision for the protocol stack implemented in a node, and quickly understands their interactions. For example, the IP module is connected to the layer 4 modules such as TCP, UDP, and those of layer 2 (Ethernet). The statistic wires are the second type of links for communication between modules. They can be traced back the statistics from one module to another, such as the size and the delay of queues of the transmitter.

Process domain defines the role of each programmable module. A module has a default main process, which can add sub-processes performing a sub-task. OPNET provides mechanisms for all processes created within a process domain to communicate with each other via a shared memory block or scheduling interrupts. The role of a module is determined by the process model, which is described as a finite state machine. Each block represents a different state, wherein the machine executes a specific code. Transitions refer to links between blocks and are determined by conditions (interruptions, variable with some value, etc.). Actions to perform are written in language *C*. OPNET provides a library of over 400 proprietary functions specific to the use of networks (creating, sending and receiving packets, extracting values in different fields of a header, etc.).

4.7.1.5.3. Network traffic model

In order to model the real-word Internet Traffic, we studied the performance of algorithms with increasing packet traffic load. According to the traffic model, we used MMPP [OKA 09] as arrival traffic process. First, this section presents briefly the traditional Poisson process. Then, we describe the MMPP and explain why we have chosen this traffic model.

– **Traditional Poisson process:**

In traditional traffic theory, the Poisson arrival process [KIN 93] was very suitable for data and voice transmission. It is a discrete probability distribution, which describes a random variable representing the number of events that occur in a time interval. In the Poisson arrivals model, there is a convenient relationship between *Poisson* (discrete) and *Exponential* (continuous) distributions. The Poisson distribution represents the number of events in a time interval. On the other hand, the exponential distribution represents time between events. The Poisson arrival model gives the following probability of k arrivals in a unit of time (1 s):

$$p_k = \frac{(\lambda)^k e^{-\lambda}}{k!} \qquad [4.43]$$

According to inter-arrival time, equation [4.44] expresses the exponential distribution of inter-arrival times:

$$f_x(t) = \lambda e^{\lambda t} \qquad [4.44]$$

The Poisson arrival process has some properties for analysis [CHE 07]:

– It is memoryless. If the previous arrival occurred T time ago, the time to the next arrival will be exponentially distributed with a mean of $\frac{1}{\lambda}$ regardless of T.

– The number of arrivals in an interval of length t has a Poisson distribution with mean λt.

– The sum of two independent Poisson arrival processes is a Poisson process with a rate of the sum of these Poisson processes. For example, the sum of two Poisson processes with rates λ_1 and λ_2 is a Poisson process with rate $\lambda_1 + \lambda_2$.

– **Markov-modulated Poisson process (MMPP):**

Since the arrival rate λ of the Poisson process is constant, it is obviously unrealistic as a traffic model. For example, with the aggregation of multiple packet speech flows, the starting and terminating times are random times, so the flow rate is not constant. MMPP has been introduced to solve this issue. The MMPP method modulates the

Poisson rate as the states of a continuous-time Markov chain. The MMPP has the memoryless property of the Poisson process and can be analyzed by Markov theory. The complexity of the analysis increases with the number of states, so this number is usually small.

Definitively, we have designed a Markov chain with three states of rate values: 1, 4 and 10 (Figure 4.20). In Figure 4.20, three states represent three rate values. The number on each transition arrow is the probability of state transition.

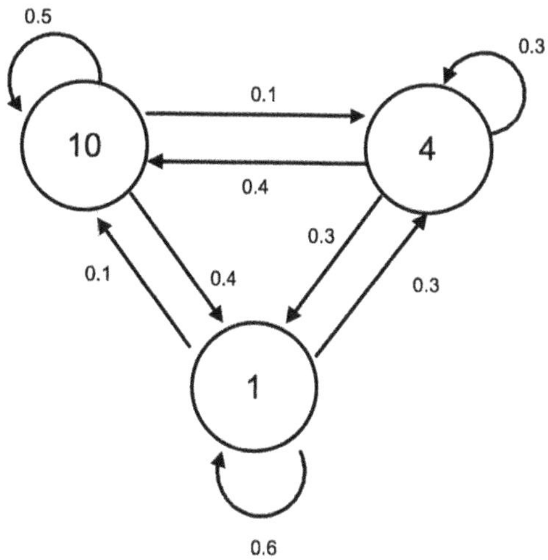

Figure 4.20. *Poisson rate modulated by Markov chain*

Applying MMPP method gives us the experiment results in Figure 4.21, which shows the switch between the states following the Markov chain in Figure 4.20. The axis y in these figures represents the number of arrival process in 1 s and the axis x is the time (in second). Taking an example of Figure 4.21a, it makes a switch from state of $\lambda = 1$ to state of $\lambda = 4$ at the 300th s. Subsequently, at the 650th s, it returns to the state of $\lambda = 1$. Then, during about 120 s from the 780th s to the 900th s, the state is $\lambda = 10$. The switch between the states follows the Markov chain in Figure 4.20. The x-axis follows the exponential distribution given in equation [4.44] with the λ being the value of current λ.

In this book, we used the MMPP model to implement the arrival process for the user request model. Actually, using the MMPP model gives a traffic model with different load level for network system. In fact, the higher the value of λ is, the more the network system is loaded.

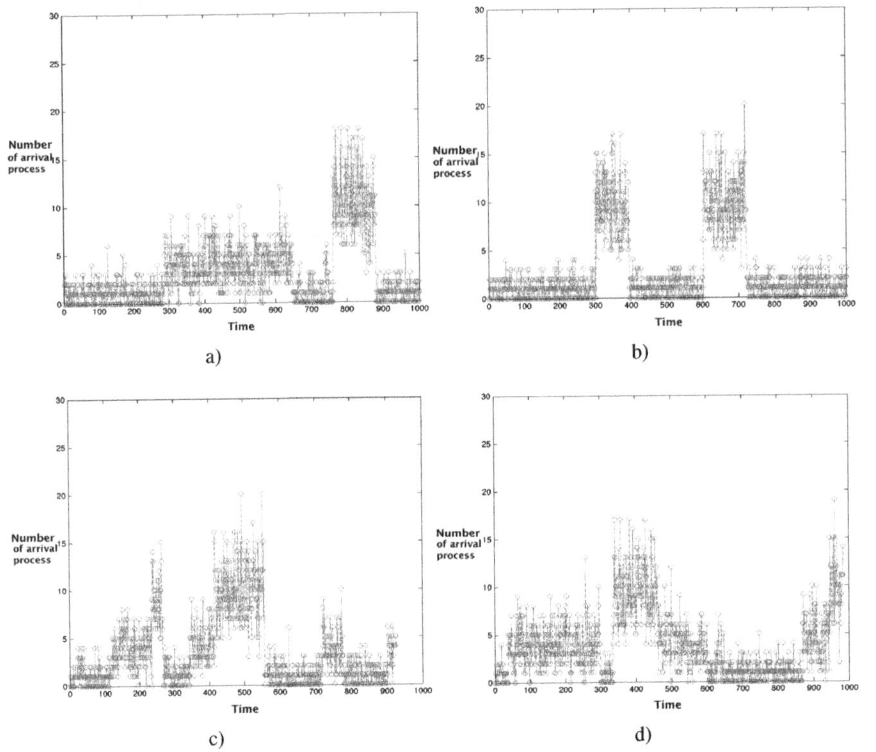

Figure 4.21. *Markov-modulated Poisson process*

4.7.2. *Experimental setup*

To validate the results of QQAR, we compare our approach with three kinds of algorithm:

– *Those based on the distance-vector (DV) algorithm.* The DV algorithm builds routing tables where no router has the overall vision of the network. The distribution of paths is gradual. The term "distance-vector" is that the protocol manipulates vectors (arrays) of distances to other nodes in the network. The "distance" is the number of hops to reach the neighboring routers. In this algorithm, routers can maintain the optimal route by storing the address of the next router in the routing table so that the number of hops to reach the destination is minimal. The routes are updated periodically (e.g. the updated period of routing internet protocol (RIP) protocol is 30 s). The DV algorithm uses the Bellman–Ford algorithm to find paths. It is a dynamic programming algorithm for finding shortest paths from a given source node in a weighted directed graph. Unlike Dijkstra's algorithm, which can only be used when all edges have non-negative weights, the Bellman–Ford algorithm allows the presence of some arcs of negative weight and detects the existence of a circuit absorbent, that is to say the total weight of negative, accessible from the top source.

– *Those based on the link-state algorithm: short path first (SPF)*. In this algorithm, each router establishes relations with its neighbors by sending hello messages. Each router then forwards the list of networks it is directly connected to by messages (LSA–link-state advertisements) to gradually spread to all routers in the network. The set of LSAs forms the database links (link-state database (LSDB)), which is identical for all participating routers. Each router then uses Dijkstra's algorithm to determine the shortest path to each network known in the LSDB.

– *Those based on standard the optimal QoS multipath routing (SOMR) algorithm*. In this algorithm, routing is based on finding the K-best optimal paths and uses a composite function to optimize delay and link cost criteria simultaneously.

While we can use static routing for small networks like local area networks (LAN), larger (or huge) networks like the Internet have complex topologies that need adaptive routing protocols, which attempt to solve this problem by constructing routing tables automatically. Three adaptive routing algorithms described above are representative of today's widely used routing systems. This motivates us to use them to compare with our proposed routing algorithm.

The comparison in our experiment is based on the average MOS score, PDR and NRO that represents the QoE value. For the video streaming, we used two types of video: the Standard Definition Video (SD Video) and the High-Definition Video (HD Video). The resolution of SD Video is 768×576, while the resolution of HD Video is $1{,}280 \times 720$ pixels (720p). We use the HD Video with large volume files to test networks with lots of loaded flows. In other words, we intend to test our network in different load levels. The first set of experiments was conducted by using the SD video files for video streaming. Then, we use the HD Video files to test our approach in a stressed network. Finally, we stressed the network by adding some data flow with different load levels.

We observe the average MOS score at 10 access points and the experiment has been relaunched 10 times as a cross-validation method. The average MOS value is calculated with sliding windows of 1 h. We also present the confidence interval of 95%. The duration time is 45 h.

4.7.3. *Average MOS score*

Figures 4.22, 4.23 and 4.24 illustrate the experimental result of average MOS scores of four algorithms in three network topologies: *the 3AreasNet*, the 6×6*GridNet* and *the NTTnet*, respectively. As mentioned previously, for each topology, we implement two types of video: the SD Video and the HD Video. For these experiments, we collect the MOS score every 5 h.

In observing Figure 4.22a and 4.22b, we see that after the first 5 h, the four protocols obtain the stable MOS score. These stable values are 4.1, 3.3, 2.95 and 2.6 for QQAR, SOMR, SPF and DV, respectively (Table 4.3).

	QQAR	DV	SOMR	SPF
MOS score	4.1	2.6	3.3	2.95
Improvement of QQAR	–	157.7%	124.2%	138.98%

Table 4.3. *MOS score for SD Video streaming of 3AreasNet topology*

According to the effect of network load on the results, video streaming with HD Video depresses QoE value of four protocols with the decrease from 4.1 down to 3.65, 3.3 to 3.2, 2.95 to 2.8 and 2.6 to 2.45 for QQAR, SOMR, SPF and DV, respectively (Table 4.4).

	QQAR	DV	SOMR	SPF
MOS score	3.65	2.45	3.2	2.8
Improvement of QQAR	–	148.97%	114.06%	130.35%

Table 4.4. *MOS score for HD Video streaming of 3AreasNet topology*

Even the HD Video streaming decreases the QoE value of protocols, the obtained MOS score of QQAR is better than the three other ones with the improvement of 148.97%, 114.06% and 130.35% for DV, SOMR and SPF, respectively.

	QQAR	DV	SOMR	SPF
MOS score	4.1	3.2	3.5	3.35
Improvement of QQAR	–	128.12%	117.14%	122.39%

Table 4.5. *MOS score for SD Video streaming of* 6×6*GridNet topology*

Similarly, in the 6×6GridNet topology (Figure 4.23), the four protocols obtain the stable MOS scores: 4.1, 3.2, 3.5 and 3.35 for QQAR, DV, SOMR and SPF, respectively. The HD Video streaming lowers the MOS scores for several approaches: 4.1 down to 3.9, 3.2 to 3.15, 3.35 to 3.3 for QQAR, DV and SPF, respectively. This decrease of MOS score is negligible. The SOMR is not affected by the increase of load.

We now analyze the results by using NTTnet topology (Figure 4.24). According to the SD Video streaming, the average MOS scores of QQAR, SOMR, SPF and DV are 4.1, 3.25, 3.05 and 2.3, respectively (Table 4.7). For the HD Video streaming, the QQAR, SOMR, SPF and DV obtain, respectively, MOS scores of 3.9, 2.9, 2.8 and 2.3.

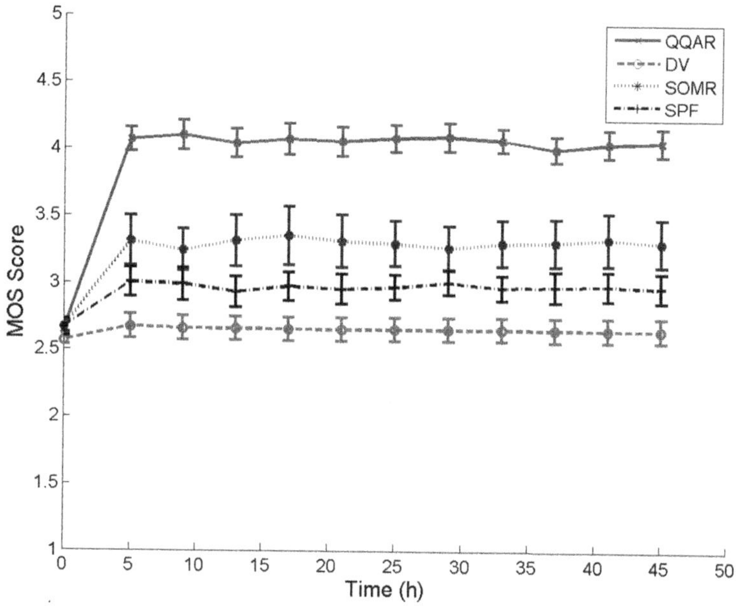

a) Using SD Video

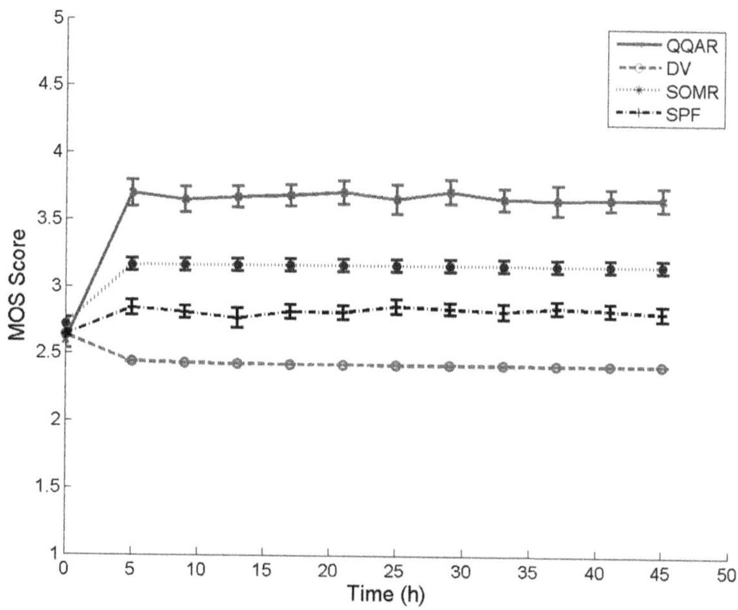

b) Using HD Video

Figure 4.22. *MOS score for 3AreasNet topology*

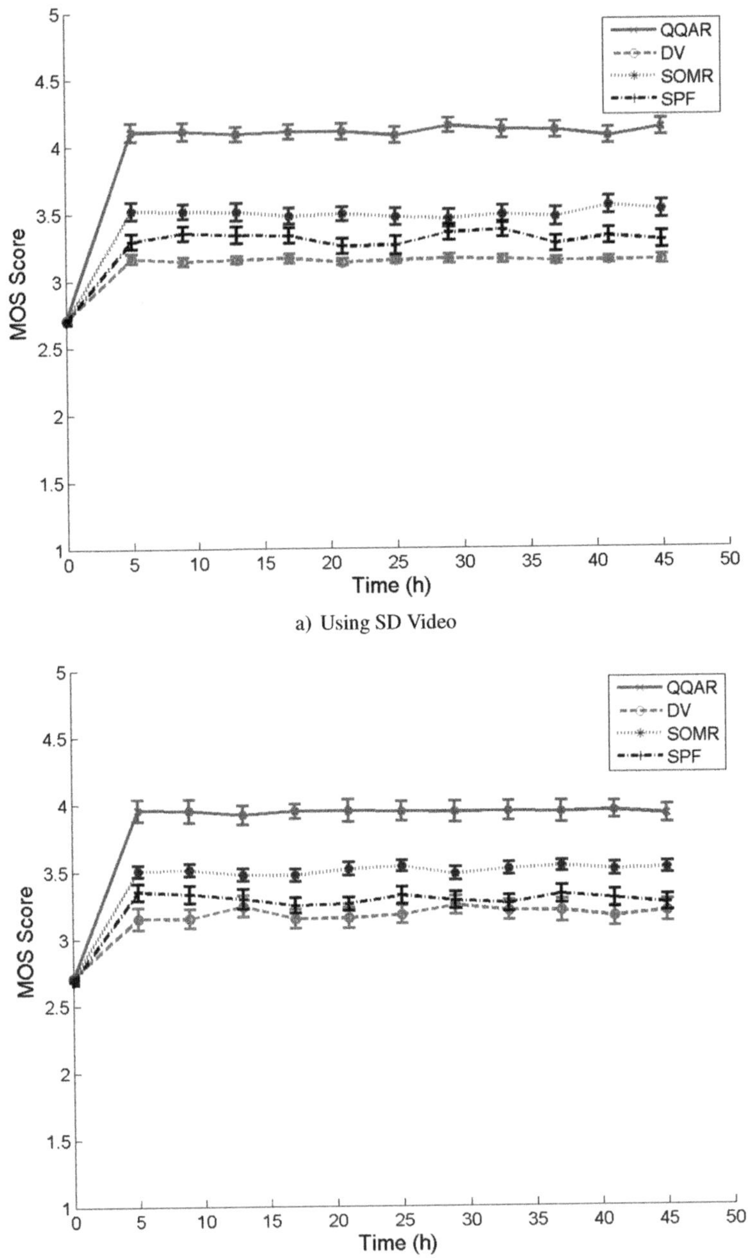

a) Using SD Video

b) Using HD Video

Figure 4.23. *MOS score for* 6×6*GridNet topology*

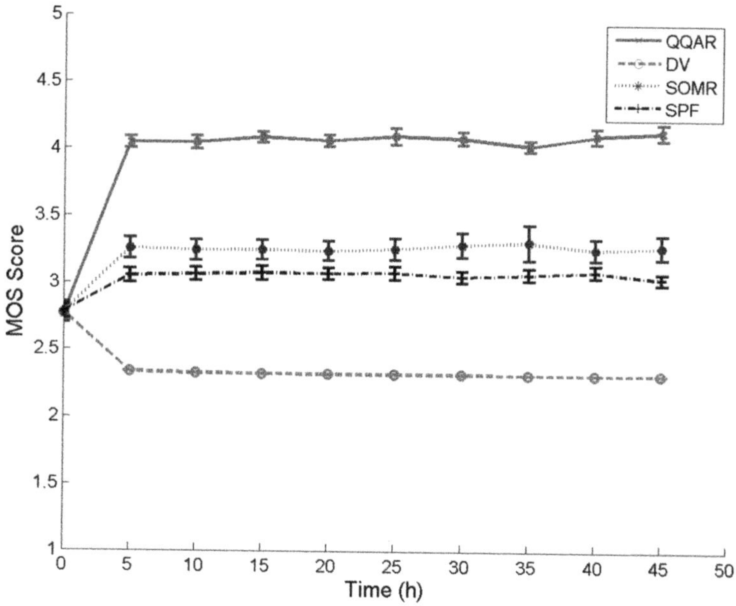

a) Using SD Video

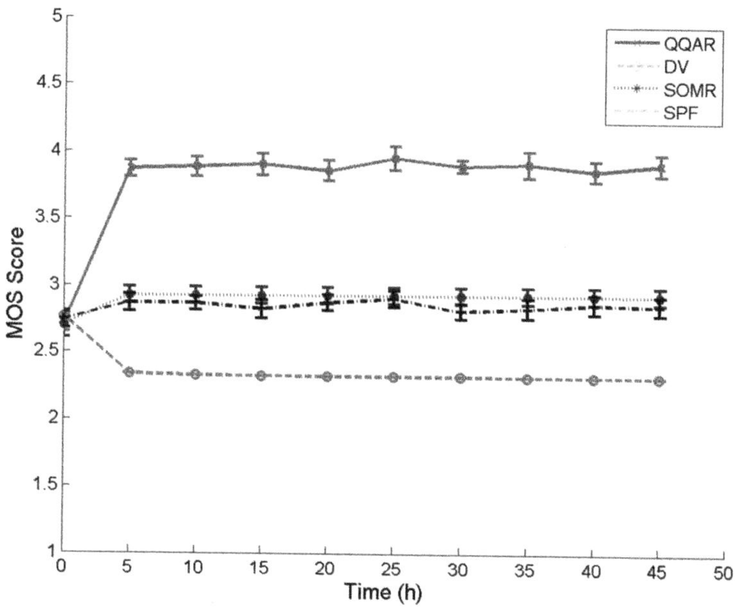

b) Using HD Video

Figure 4.24. *MOS score for NTTnet topology*

	QQAR	DV	SOMR	SPF
MOS score	3.9	3.15	3.5	3.3
Improvement of QQAR	–	123.8%	111.42%	118.18%

Table 4.6. *MOS score for HD Video streaming of* 6×6*GridNet topology*

	QQAR	DV	SOMR	SPF
MOS score	4.1	2.3	3.25	3.05
Improvement of QQAR	–	178.26%	126.15%	134.42%

Table 4.7. *MOS score for SD Video streaming of NTT topology*

	QQAR	DV	SOMR	SPF
MOS score	3.9	2.3	2.9	2.8
Improvement of QQAR	–	169.56%	134.48%	139.28%

Table 4.8. *MOS score for HD Video streaming of NTT topology*

We have the same conclusion as the previous topology: the obtained MOS score of QQAR is better than the three others in both types of video streaming.

In three networks: *3AreasNet*, 6×6*GridNet* and *NTTnet* (Figures 4.22, 4.23 and 4.24), the algorithm QQAR gives better QoE value than the three others. That is explained by the following reasons:

– While the other protocols are based on Dijkstra's algorithm, which finds the shortest paths in terms of a cost function, the QQAR algorithm is based on the Q-learning algorithm of RL, which can learn the environment changes and the obtained feedback in order to be increasingly adaptive.

– QQAR is based on a nonlinear combination of QoS parameters. More precisely, we used four network criteria: *delay*, *jitter*, *loss rate* and *conditional loss rate*. In addition, the SOMR used a composite linear function to optimize delay and link cost criteria. As mentioned previously, the user perception does not depend linearly on the technical parameters. So, a linear combination of QoS parameters is not suited to be the cost function to optimize.

– As proven in section 4.6.3, QQAR is able to detect exactly where the network issues come from without any broadcast mechanism.

According to the DV algorithm, it is the most widely accepted routing protocol. It is straightforward in broadcasting their entire current routing database periodically, typically every 30 s. The name "distance-vector" means that there is a distance (a cost) and a vector (a direction) for each destination. The DV algorithm works well for small, stable networks. This explains that it obtains a better QoE result in 3AreasNet and 6×6grid topology than NTTnet topology. In addition, there are some drawbacks of DV algorithms:

– for this algorithm, the best path is the path having the least number of node hops. However, the routing path having the lowest hop-count is not enough to guarantee good service quality;

– there is no protection from routing loops;

– it uses a limited hop-count (usually 15 s), which is an important reason for making the DV algorithm unsuitable for large networks.

These arguments explain the worst QoE results of DV compared with the three other ones.

For SPF algorithm, it is considered as a link-state algorithm. The SPF is better suited than DV for large and dynamic networks. This explains the fact that SPF obtain better results than the DV algorithm. However, in a large network, a link change can lead to a flood of thousands of link-state messages that propagate across the entire network. In each router, the database storing these messages can be quickly overloaded. Each time there is a network change, routers must redetermine new paths. Fortunately, SPF is designed to resolve this issue by devising the SPF area. SPF areas are subdivisions of an entire network. More precisely, in the 3AreasNet topology (Figure 4.19a), SPF configures by dividing the network into three separated areas. In the 6×6GridNet topology (Figure 4.19b), the network is divided into two separated parts. SPF routers in one area do no exchange topology updates with the routers in the other areas. This reduces the update message(s) number of flows through the network. This improvement is shown clearly in our convergence time studies.

According to the SOMR algorithm, it is useful for finding multiple paths between a source and a destination in a single discovery. This multiple paths-based algorithm can be used to react to the dynamic and unpredictable topology change in networks. For example, when the shortest path found in using Dijkstra's algorithm is not available for some reason, it is necessary to find the second shortest path. These paths are considered as k-shortest paths. This advantage explains the good MOS scores of the previous experiments: 3.3–3.2 in 3AreasNet, 3.5 in 6×6GridNet and 3.25–2.9 in NTTnet. The only drawback of this kind of algorithm is the time taken to calculate the k-shortest path. This algorithm is used several times to calculate the k-shortest path with $k > 1$. Because of this computational complexity, the SOMR obtains a lower quality than QQAR. In addition, the considered cost of SOMR is a linear combination of network parameters.However, the QoE depends not only on technical criteria but also on the human perception, which is an important subjective factor that we have considered in our proposals. More precisely, in our QoE measurement method, we determine the QoE by using a hybrid way that combines objective and subjective methods.

4.7.4. *Convergence time*

We now discuss the convergence time of these approaches. Network convergence is a notion that is used under various interpretations. In a routing system, every participating router will attempt to exchange information about the network topology. Network convergence is defined as the process of synchronizing network forwarding tables from the beginning (the initialization phase) or after a topology change. When there are no more forwarding tables changing for "some reasonable" amount of time, the network is said "converged". For the learning routing algorithms like QQAR, the convergence time refers to the initialization phase for exploring the network environment. After this phase, such an algorithm becomes stable.

The three Figures 4.25, 4.26 and 4.27 show the obtained MOS scores of four algorithms in the first 5 min of simulation. The MOS scores are taken every 10 s. We consider three cases: for 3AreasNet (Figure 4.25), 6 × 6GridNet (Figure 4.26) and NTTnet (Figure 4.27) topologies. The goal is to study the convergence time of each approach in the initialization phase. All of the four algorithms have a common feature: in the initialization period, they have no information about the system. So, during this period, they try to explore the network. We call this period the *initialization phase*.

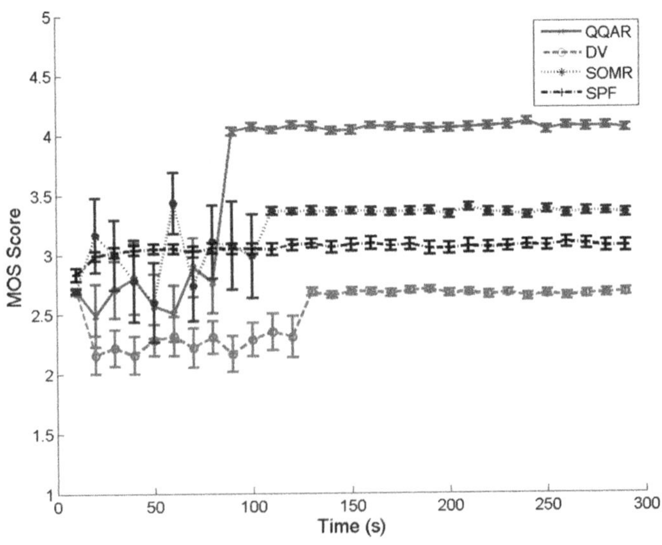

Figure 4.25. *Initialization phase for 3AreasNet topology*

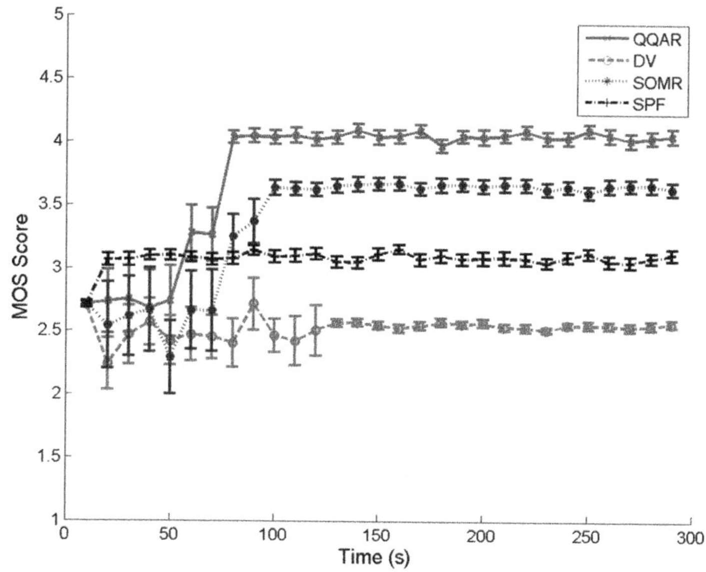

Figure 4.26. *Initialization phase for* 6×6*GridNet topology*

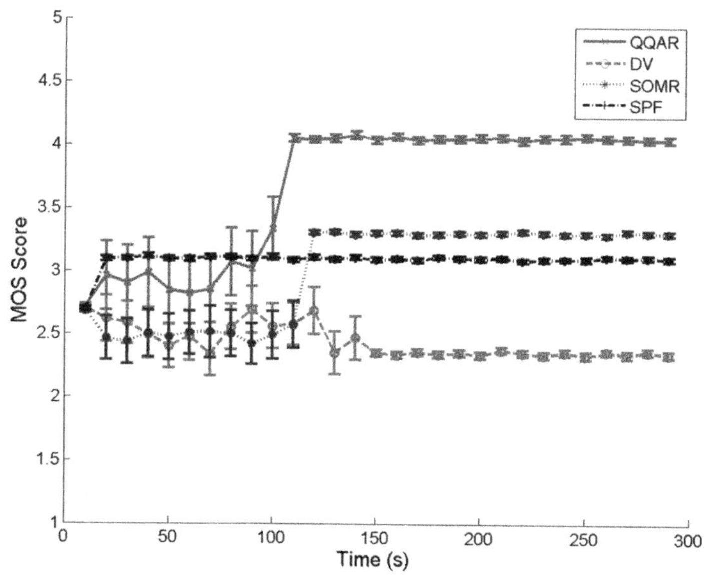

Figure 4.27. *Initialization phase for NTTnet topology*

As shown in Figure 4.26, in the 6 × 6GridNet topology, the convergence times are 20, 90, 110 and 130 s, respectively, for SPF, QQAR, SOMR and DV. For the 3AreasNet network, the convergence times are 20, 90, 110 and 130 s for SPF, QQAR, SOMR and DV, respectively. For the NTTnet, the convergence times are 20, 110, 120 and 150 s for SPF, QQAR, SOMR and DV, respectively (Table 4.9). The larger the network is, the more time it takes for convergence.

	QQAR	DV	SOMR	SPF
Convergence time for 3AreasNet (s)	90	130	110	20
Convergence time for 6 × 6GridNet (s)	80	130	100	20
Convergence time for NTTnet (s)	110	150	120	20

Table 4.9. *Convergence time*

According to the DV algorithm, it is considered as a slow convergence routing algorithm due to the slow propagation of update messages. The DV routers periodically broadcast their routing database every 30 s. Especially, in large or slow link networks, some routers may still publicize a disappeared route. This explains the long convergence time of the DV algorithm shown in Table 4.9. It obtains the stable QoE value after 130 s in 3AreasNet and 6 × 6GridNet network and 150 s in NTTnet network. DV convergence is the slowest one among the four algorithms.

In contrast to the DV, SPF algorithm converges quickly and accurately. In both network topologies, the small one (3AreasNet and 6 × 6GridNet) and the large one (NTTnet), SPF works very well with the obtained result shown in Table 4.9. It archives an impressive result of 20 s of convergence. As explained above, the SPF algorithm divides 3AreasNet into three separate areas and 6 × 6GridNet into two areas. The routers within one area do not exchange topology updates with the ones in the other areas, which reduces the number of update messages to alleviate the network load. So, these advantages make SPF the fastest algorithm in terms of convergence time among the four algorithms.

For the QQAR algorithm, the routers do not have knowledge of entire network. Each QQAR router has a routing table containing Q-values of their neighbors. The exchange phase between each router and their neighbors takes time for the convergence phase, especially in a large network like NTTnet. This explains the obtained results in Table 4.9: 80 and 90 s in 6 × 6GridNet and 3AreasNet, but 110 s in NTTnet. In fact, 110 s is still an acceptable duration for convergence time of a routing protocol.

The SOMR-based routing uses the diffusing update algorithm (DUAL) [GAR 93] to obtain loop-freedom at every instant throughout a route computation, while the DV algorithm meets temporary micro-loops during the time of re-convergence in the topology due to inconsistency of forwarding tables of different routers. Therefore, as shown in Table 4.9, SOMR gives better convergence time results than the DV algorithm. More precisely, these are 100, 110 and 120 s in 6 × 6GridNet, 3AreasNet and NTTnet, respectively.

4.7.5. *Capacity of convergence and fault tolerance*

We have also tested the algorithm's capacity of convergence as well as fault tolerance when a change of topology occurred in the network. To implement this situation, we broke down the link *R*18-*R*19 of the 6 × 6GridNet topology (Figure 4.28) at the 20th h. After this change of topology, the two areas (the left and the right) of the 6 × 6GridNet network are connected by only one link: *R*16-*R*25.

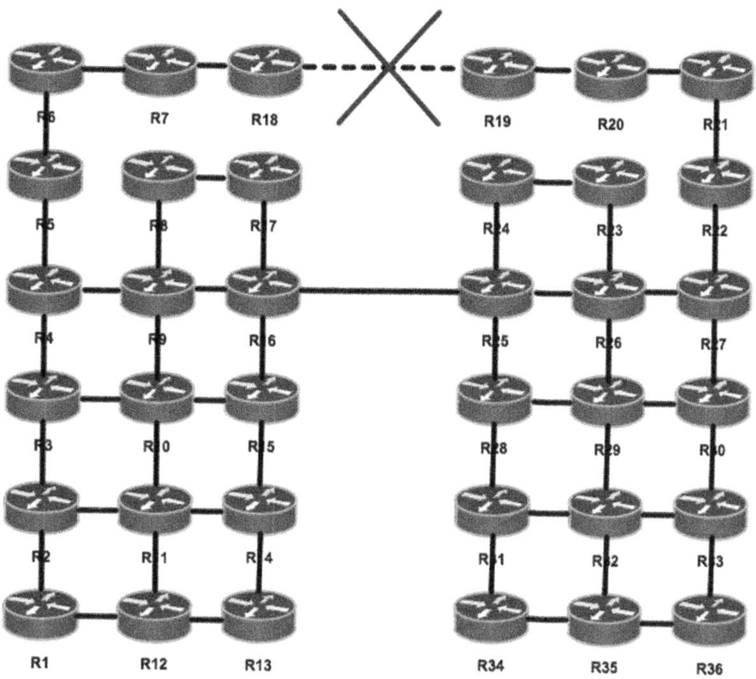

Figure 4.28. 6 × 6*GridNet topology after breaking down the R18-R19 link*

As shown in Figure 4.29, after the 7200th s where the link *R*18-*R*19 was broken, the MOS scores of the four algorithms decreased to a value of 1.7. According to the convergence time (Table 4.10), the three algorithms QQAR, SOMR and SPF take the same time as the convergence time for initialization phase (Figure 4.26). However, even after the convergence phase, the re-obtained MOS scores of these algorithms are lower than the ones before the topology change. More precisely, the QoE value decreases from 4 to 3.7 for QQAR, 3.6 and 3.4 for SOMR, 3-2.8 for SPF and 2.55

to 2.4 for DV. This fact can be explained as follows. After the topology change, the data exchange between two areas of 6×6GridNet network is realized only on one link: the link $R16$-$R25$. Therefore, the congestion issue of this link caused the quality degradation. One more notable comment, the DV algorithm has no convergence phase in this context because even before the topology change, it used just the link $R16$-$R25$ to exchange data between two areas. So, the disappearance of the link $R18$-$R19$ does not have much influence on the DV algorithm result.

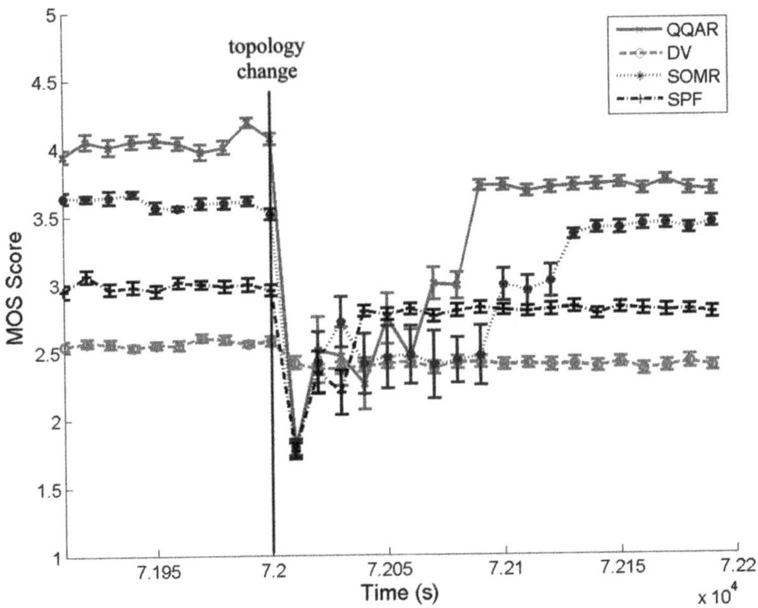

Figure 4.29. *MOS score after topology change*

	QQAR	DV	SOMR	SPF
Convergence time (s)	80	–	130	20

Table 4.10. *Convergence time for a topology change*

Figure 4.30 shows us the average of these four algorithms in different load levels formulated in [4.37]: from 10% to 70%. We can see that the more the system is loaded, the more the average MOS score decreases. However, at any load level, the average MOS score of QQAR is higher than the three other algorithms. For a load level of 10%, the MOS score of QQAR is 4.1 (note that 4 represents a good quality in MOS score range). Regarding the three other protocols, the maximum value obtained by SOMR is just 3.3 with a load level of 10%.

QQAR shows a better end-to-end QoE result than the three other algorithms in any load level and in the same network context. So, with our approach, despite network environment changes, we can maintain a better QoE than traditional algorithms. Thus, QQAR is able to adapt its decisions rapidly in response to changes in network dynamics.

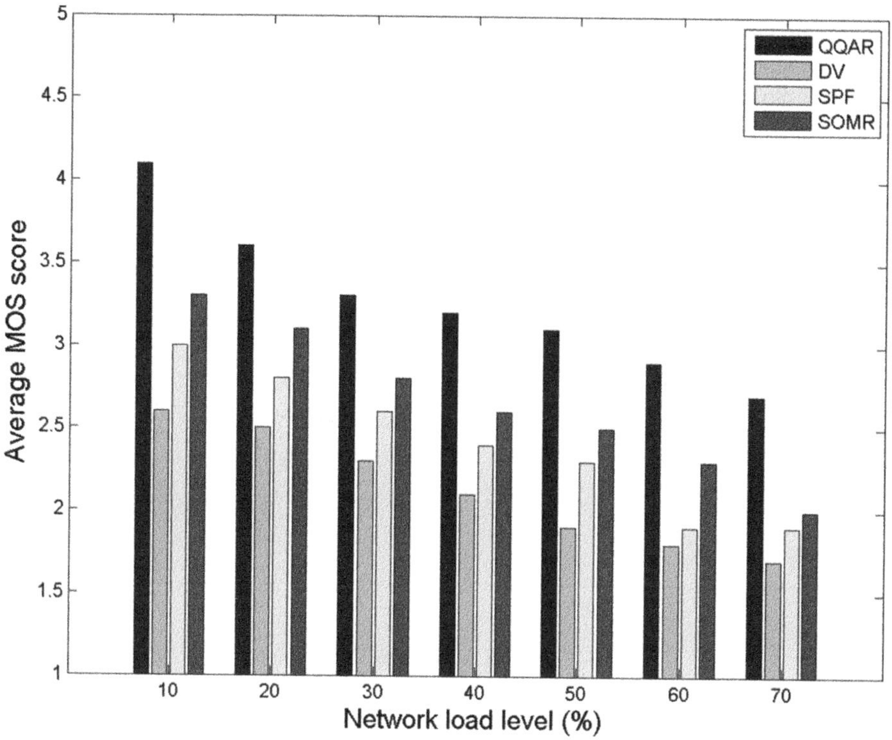

Figure 4.30. *User perception in different load levels*

4.7.6. *Control overheads*

Our experimental works also consist of analyzing overheads of these protocols. The type of overhead we observe is the control overhead that is determined by the proportion of control packet number as compared to the total number of packets emitted. To monitor the overhead value, we have varied the node number by adding routers to the network system. So the observed node numbers are 38, 50, 60, 70, 80. The obtained results are shown in Figure 4.31.

We can see this the control overhead of our approach is constant (50%). This is explained by the equal number of control packets and data packets in QQAR. Each generated data packet leads to an acknowledgment packet generated by the destination node. The control packet rates of DV, SPF and SOMR are respectively 0.03, 0.4 and 0.1 (packet/s). This order explains the control overhead order in Figure 4.31. While the SPF algorithm has the highest value because of the highest control packet rate (0.4 packet/s) with multiple types of packets such as Hello packet, Link State (LS) Acknowledgment packet, LS Update packet and LS State request packet, the DV algorithm has the smallest overhead value with a control packet rate value of 0.03. We can also see that the higher the number of nodes, the higher the overhead. So, with a stable overhead, our approach is more scalable than these other three.

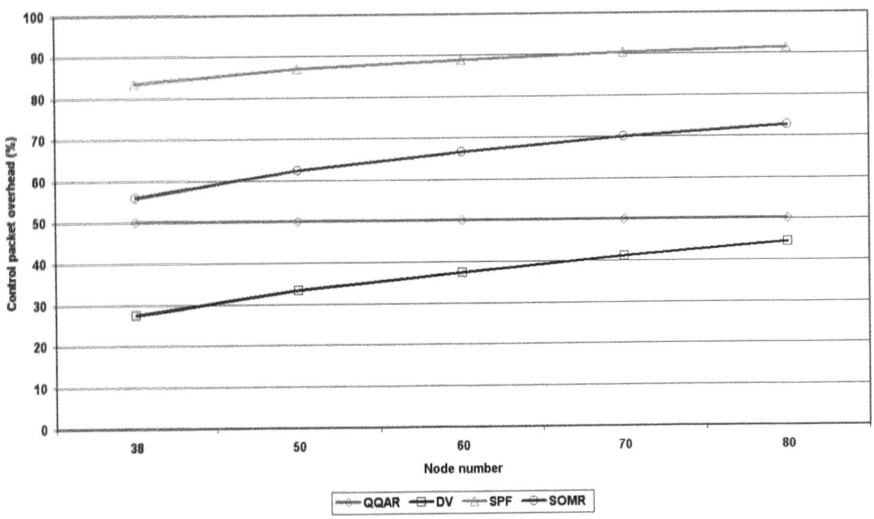

Figure 4.31. *Control overhead*

4.7.7. *Packet delivery ratio*

According to the packet delivery issue, Figure 4.32 shows the PDR of the algorithms. PDR is the ratio of the number of packets received by the destination (client nodes) to the total number of sent packets at server nodes. This is calculated by using equation [4.40]. Generally, the more the network is loaded, the more PDR decreases. The QQAR gives a good result in keeping the ratio at 100% until the network load level of 30%. The minimal value obtained is 87% at 70% of load level. The PDR of the three other algorithms decreases with the increase of network load level. When network load level reaches 70%, the PDRs are 87%, 70%, 61% and 49% for QQAR, SOMR, SPF and DV, respectively. So, the QQAR obtains the best result in terms of delivery ratio.

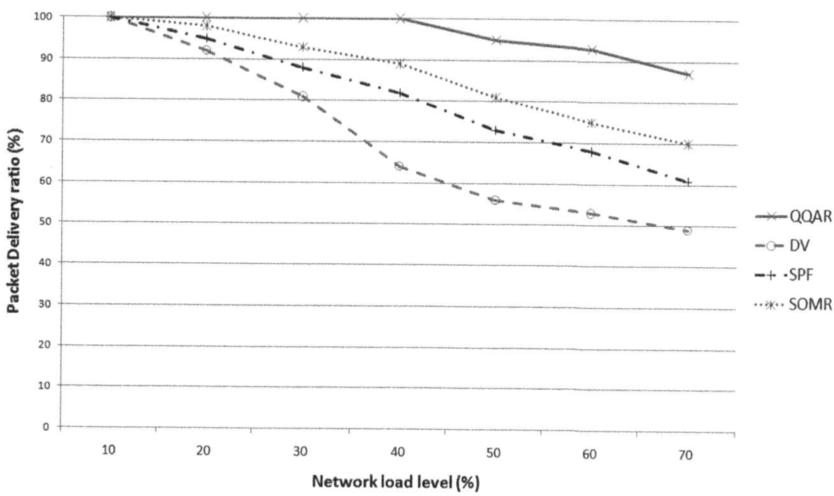

Figure 4.32. *Packet delivery ratio*

4.8. Conclusion

In this chapter, we present an example of routing systems based on the QoE notion. We have integrated the QoE measurement into the routing paradigm for an adaptive and evolutionary system, called QQAR. Definitively, the latter is based on Q-learning, an algorithm of RL, and used QoE feedback as a reward in the learning model. The goal is to guarantee end-to-end QoE of a routing system. A simulation conducted on a large, real test bed demonstrates that this kind of approach yields significant QoE evaluation improvements over other traditional approaches. More precisely, the simulation results showed that the QQAR algorithm performs better than the other approaches in terms of the following: (1) MOS for different network loads; (2) convergence time in initialization phase; and (3) control overhead.

5

User-Driven Server Selection Algorithm for CDN Architecture

This chapter will present the proposed server selection method that is inspired by the bio-inspired multi-armed bandit problem for the meta-routing layer of a content distribution network architecture. The proposal considers the Quality of Experience as the reward regulation parameter in the control process that is able to take into account the exploration–exploitation trade-off.

5.1. Introduction

The end-to-end QoE model is efficient and indispensable in each of today's network systems. It carries the user experience to guarantee the satisfaction of end users instead of focusing only on technical parameters such as current Quality of Service (QoS) systems. In fact, as proven previously, satisfying end users is the new trend of development for the Internet as well as emerging high-quality applications. However, the end-to-end QoE model will be abstract if it is not applied to any real network architecture. In this book, we decided to apply this model to a CDN architecture, which has been introduced to improve the delivered service quality.

As discussed in Chapter 3, CDN [YIN 10] is considered as the user-centered solution in moving content from the origin servers to the servers close to end users. This approach is used to improve the scalability of a network service due to the fact that it is able to provide services to a large number of clients with a reasonable service quality. Actually, the user requests are transferred to replica servers instead of being sent to the origin server. That helps the system to avoid congestion as well as overload issues at the core network and server's side. In addition, this cache-related approach minimizes the delay time. Theoretically, a CDN architecture has two main layers: the routing layer and the meta-routing layer. The second layer is composed of several modules such as *server placement, cache organization* and *server selection.* The two first modules belong to the preparation-related phase. More precisely, the server placement module tries to place the replica servers in an optimized way to minimize the delivery delay and the bandwidth consumption. Providers use the cache organization module to organize the content stored in replica servers in order to guarantee the availability, freshness and reliability of content. Besides these two preparation-related modules, we focus on the server selection module, which plays an important role in launching the operation phase of a CDN. The fundamental objective of server selection is obviously to offer better performance than that of the origin server. Another added value of this selection process is the lowering of the costs of network resources. It is not easy to choose an appropriate server to provide service to users. The appropriate server may not be the closest one in terms of hop-count or end-to-end delay, or the least loaded server. The best server is the one that makes end users satisfied when using the provided service. So, the server selection process plays a key role in the decisive success of a CDN. We now analyze some related research on server selection methods in the context of CDN. Then, we explain our motivation to develop a server selection scheme based on a MAB formalization.

Zhou *et al.* [ZHO 08] present a server selection method that is based on load information of replica servers to choose the appropriate server. This method uses a reliable server pooling framework named RSerPool. Unfortunately, this approach is not always feasible because the accessibility of CDN servers is not always available. In fact, server polling or checking with a high frequency increases the server load.

Han *et al.* [HAN 09] propose a server selection method that uses round trip times (RTTs) between client and servers to choose the server that has the minimum RTT to the client. The RTTs are measured by passive-based probing. The drawback of this proposal is that probing may increase traffic overhead of the network.

Chellouche *et al.* [CHE 10] use an anycasting technique to realize a server selection process. This mechanism chooses the appropriate server among a server set in an anycast group based on network metrics. However, using an anycasting system is not easy and common. Even if it is applicable to IPv6 infrastructure (RFC 1546), the IPv4 network infrastructure does not support this kind of delivering operation because there is not an address range set aside for anycast use on the public IPv4 Internet (presented in IETF RFC 3330).

Wendell *et al.* [WEN 10] propose a distributed system, called DONAR, that uses an algorithm to consider both client performance (e.g. sudden changes in client request rates) and server load. This system uses mapping nodes for server selection. This algorithm can maintain a weighted split of requests to a client's replicas. This approach uses a weighted function for decision-making. So, choosing the appropriate weighted vector is a problem of this approach because we cannot obtain an optimized solution.

In several projects such as the RingServer Project in Japan [RIN 09], the round robin mechanism has also been used for server selection policy. However, since each server is selected by a classic round robin algorithm, both the network criteria and server load are ignored. This simple policy is not applicable in a dynamic network system such as CDN.

Guo *et al.* [GUO 02] design and evaluate server selection techniques with a fixed number of channels for replicated video-on-demand servers. The authors propose several heuristic techniques, to outperform a basic policy, that always route requests to the closest server. For the same purpose, Pierre *et al.* [PIE 06] propose Globule. The closest server in terms of network proximity is always chosen. However, the closest server is not always the server that gives the best quality. Similar to above works, Almeida *et al.* [ALM 04] try to minimize the objective function of a weighted sum of network and server bandwidth usage to resolve the server selection problem.

At this point of our work, a fundamental point must be made: all of the approaches presented above consider only the QoS criteria, such as network and server parameters, as the measure of performance to optimize, but not the QoE that represents the actual end user perception. The assumption made behind this decision is that a good QoS measurement strictly induces a good QoE performance. However, this hypothesis is not always correct because the data are delivered to end users through the network in different contexts. So, the subjective end user context has to be considered carefully, or at least not neglected. In addition, these approaches do not take into account the exploitation–exploration trade-off (i.e. finding a balance between uncharted knowledge and current knowledge) that must be taken into account in any online decision-making task such as in a CDN environment. Consequently, these above problems motivate us to propose a novel server selection algorithm, which takes into consideration the end user's perception as well as the exploitation–exploration trade-off. To this purpose, we have to use a selection algorithm based on a learning method.

Taking an example of a learning problem, we have to repeatedly choose among n different options (*actions*). After each choice, we receive a numerical reward of a stationary probability distribution dependent on the chosen action. The goal is to maximize the expected total reward over some time period. This is the form of the MAB problem that we applied to formalize the server selection issue in a CDN.

This section presents the MAB formalization and explains why we chose it to formalize our selection algorithm. Then, we focus on the proposed server selection algorithm.

5.2. Multi-armed bandit formalization

MAB [VER 05, ANT 08] takes its name from a traditional slot machine (one-armed bandit). With the motivating applications in statistics, multiple levers of this machine are considered. After pulling a lever, the latter provides a reward drawn from a distribution associated with that specific lever. The goal of the gambler is to maximize his total rewards earned in a series of trials. Indeed, many learning methods and optimizing problems are formalized in this way.

In this section, we briefly present the notion and general formulation of the MAB and also discuss some common policies for the MAB problem.

5.2.1. *MAB paradigm*

We chose the MAB formalization for our proposal due to the following reasons:

– MAB formalization does not make any assumption concerning the reward function. This approach always carries the exploration and exploitation trade-off dilemma for any reward function. The reward function we take into account is the mean opinion score (MOS), the evaluated QoE value.

– Unlike other approaches of QoS optimization problems, MAB makes no heuristic. While other classic approaches try to optimize a concrete heuristic such as end-to-end delay, and hop-count, MAB explicitly minimizes the regret ρ (equation [5.2]).

– Among the approaches capable of solving the MDP, there is only the MAB approach that needs only one state. The future action is chosen based on the obtained reward and this unique state. This property is suitable for the server selection issue we are working on because we only know the current state of each server.

In the MAB problem, the expected reward given by the selected action is called the *value* of this action. The MAB would be easily solved if we knew the value of each action; we would always choose the action with highest value. In reality, we do not know the action values, although we may have estimates. In fact, among actions with different action values, there is at least one action at any time whose estimated value is the greatest. This action is called a *greedy action*. Selecting the greedy action means *exploiting* the current knowledge of the value of the actions. However, selecting one of the non-greedy actions means *exploring* because this enables us to improve the estimate of the non-greedy action's value. To compare exploitation and exploration, while exploitation is used to maximize expected reward on one choice, exploration is used to produce greater total reward in the long run. For example, we know with certainty the greedy action's value, while other actions are evaluated to be nearly as good but with substantial uncertainty. However, at a certain point, the uncertainty means that one of other actions probably becomes better than the greedy action, but we do not know which. It is clearly better to explore the non-greedy actions and determine which of them are better than the greedy actions. Rewards may be lower in the short run, during the exploration phase, but higher in the long run after we have determined the better actions to exploit them. Since we cannot both explore and exploit in a single selection action, we call that *conflict between exploration and exploitation.*

There are some algorithms for evaluating the value of actions and for using the estimates to select the action. One simple way to estimate this is by averaging the rewards received if the action was selected. So, after t decisions, the action value of action a is estimated after n_a times as follows:

$$Q_t(a) = \frac{r_1 + r_2 + ... + r_{n_a}}{n_a} \qquad [5.1]$$

where $Q_t(a)$ is the estimated value after t choices. As $n_a \to \infty$, $Q_t(a)$ converges to $Q * (a)$.

The above method, called *sample-average,* is one way to evaluate action values, but not the best one. For example, the algorithm UCB1 that we used in our approach has the estimate function, called the upper confidence bound (UCB) index, of $\hat{r}_a + \sqrt{\frac{2\ln(t)}{n_a}}$ where the $\hat{r}_a$ is the estimate for the true expected rewards.

In many real-world contexts, we make decisions to maximize the expected numerical reward. The chosen actions are used not only to bring in more reward but also to discover new knowledge that could help to improve future decisions. Such situations include adaptive server selection efforts for improving the service quality in a CDN system. All decision-making systems face a particular problem of balancing reward maximization based on the already acquired knowledge and attempting new actions to further increase knowledge. That is known as the exploitation–exploration trade-off in reinforcement learning.

The MAB was introduced by Robbins [ROB 52] to solve this general problem. The main difference with a traditional slot machine (one-armed bandit) is that the MAB has more than one lever. Pulling a lever provides a reward drawn from a distribution associated with that specific lever. In the beginning, the gambler has no knowledge about the levers, but after some repeated trials, he can determine the most rewarding levers. In fact, the MAB problem is actually determining the best policy for the gambler.

The MAB problem is a formalization of a sequential decision-making task composed of a decision maker and fixed decision actions a_i with $i = 1..N$ $(N \geq 1)$. Each time the decision maker chooses an action a_i^t at time t, it receives a reward $r_{a^t}^t$ drawn from an unknown distribution associated with that specific action. The goal of the decision maker is to maximize the sum of rewards received through a sequence of choosing actions. So in each period, $t = 1, 2, ..., T$:

– The decision maker chooses a^t, based on its previous rewards and chosen actions $a^1, r^1_{a^1}, a^2, r^2_{a^2}, ..., a^{t-1}, r^{t-1}_{a^{t-1}}$.

– The decision maker receives only the reward $r^t_{a^t}$.

The MAB problem can be modeled by a set of real distributions $\Psi = \{\Phi_1, \Phi_2, ..., \Phi_N\}$ called arms. Each distribution Φ_i is associated with the rewards delivered by action a_i. The average values associated with these reward distributions are $\xi_1, \xi_2, ..., \xi_N$.

The regret ρ after T rounds is defined as the difference between the reward sum associated with an optimal strategy and the sum of the collected rewards:

$$\rho = T\xi^* - \sum_{t=1}^{t=T} r_t \qquad [5.2]$$

where ξ^* is the maximal reward average, $\xi^* = \max(\xi_i)$, and r_t is the reward at time t.

All of the policies or algorithms proposed for a MAB problem tend to minimize the regret per round. A strategy is called *zero-regret policy* if its average regret per round tends to zero with probability of 1 for any bandit problem when the horizon tends to infinity. With a sufficiently played round, the zero-regret policies are guaranteed to converge to an optimal policy. This formalism is a common description of a discrete action based on a reinforcement learning problem [SUT 98] with a unique state.

There are many extensions to the problem of stochastic MAB:

– *Bandit in MDP* [JAK 10]: exploration strategy "optimistic uncertainty" (based on USB) to explore a Markov decision process. In [AUE 09], the authors proposed an algorithm, called upper confidence reinforcement learning (UCRL), that achieves logarithmic expected regret: $\mathbb{E}[R^A_T] = O(\log T)$. Fix a time t. UCRL keeps an estimate of the transition probabilities ($p_t : X \times A \times X \to [0, 1]$) and the immediate rewards ($r_t : X \times A \to [0, 1]$). Furthermore, UCRL constructs confidence bounds, $c^{(p)}_t : X \times A \times X \to \mathbb{R}^+$, $c^{(r)}_t : X \times A \to \mathbb{R}^+$, for these estimates: $c^{(p)}_t(x, a, y)$ is a confidence bound for the transition probability estimate $p_t(x, a, y)$, and $c^{(r)}_t(x, a)$ is a confidence bound for the reward estimate $r_t(x, a)$. Together these specify the set of plausible transition probabilities and rewards, given the learner's experience up to time t: in particular, p is a plausible transition probability function, if it is a transition probability function, and $|p(x, a, y)p(x, a, y)| \leq c(p)(x, a, y)$ holds for all $(x, a, y) \in X \times A \times X$. Similarly, r is plausible if $r : X \times A \to [0, 1]$ and

$|r_t(x,a) - r(x,a)| \leq c_t^{(r)}(x,a)$ holds for all $(x,a) \in X \times A$. Denote the set of plausible MDPs at time t by M_t. The "optimistically best policy at time t" is $\pi_t = argmax_\pi\{\rho^\pi(M) : M \in M_t\}$.

– *Bandits with contextual information*: at each time t, there is information $x_t \in X$ and we make a decision $a_t \in A$. The reward is a function of a_t and x_t. It is comparable to a class of strategies $\pi : X \to A$.

– *Bandit with a countable number of arms* [WAN 08]: each new arm has a probability ϵ^β to be ϵ-optimal. In [WAN 08], the authors focus on the case when the number of arms is infinite (or larger than the available number of experiments). In this case, the exploration of all the arms is impossible to achieve: if no additional assumption is made, it is not easy to find a near-optimal arm. They consider a stochastic assumption on the average-reward of any new selected arm. After pulling a new arm k, its average-reward μ_k is assumed to be an independent sample of a fixed distribution. They write $v_n = O(u_n)$ when for some n_0, $C > 0$, $v_n \leq Cu_n(\log(u_n))^2$, for all $n \geq n_0$. They assume that the rewards of the arms lie in $[0,1]$. The regret bounds depend on whether $\mu^* = 1$ or $\mu^* < 1$. For $\mu^* = 1$, the algorithms are such that $\mathbb{E}R_n = O(n^{\beta/(1+\beta)})$. For $\mu^* < 1$, we have $\mathbb{E}R_n = O(n^{\beta/(1+\beta)})$, if $\beta > 1$, and $\mathbb{E}R_n = O(n^{1/2})$, if $\beta < 1$. In addition, they derive the lower bound: for any $\beta > 0$, $\mu^* \leq 1$, any algorithm satisfies $\mathbb{E}R_n \geq Cn^{\beta/(1+\beta)}$ for some $C > 0$. The authors propose an algorithm having the anytime property, which is based on an arm-increasing rule. The proposed algorithms essentially consist of pulling K different arms chosen randomly, where K is of the order $n^{\beta/2}$ if $\mu^* < 1$ and $\beta < 1$, and $n^{\beta/(1+\beta)}$ otherwise, and using a variant of the UCB algorithm on this set of K arms, which takes into account the empirical variance of the rewards. This last point is crucial to obtaining the proposed rate for $\mu^* = 1$ and $\beta < 1$.

– *Linear arms* [MER 09]: we choose an arm $x_t \in X \subset \mathbb{R}^d$. The average reward is a linear function $r_t = x_t \cdot \alpha$, where $\alpha \in \mathbb{R}^d$ is an unknown parameter. In [MER 09], the authors propose a simple model that demonstrates the benefits of exploiting the underlying structure of the rewards. They consider a bandit problem where the expected reward of each arm is a linear function of an unknown scalar, with a known prior distribution. Since the reward of each arm depends on a single random variable, the average rewards are perfectly correlated. They prove that, under certain assumptions, the cumulative Bayes risk over T periods under a greedy policy admits an $O(\log T)$ upper bound, independent of the number of arms. This approach is extended in [RUS 10] to the setting where the expected reward of each arm depends linearly on a multivariate random vector $Z \in \mathbb{R}^r$. The authors in [RUS 10] concentrate on the case where $r \geq 2$, which is fundamentally different from the model in [MER 09] due to the fact that the average rewards now depend on more than one random variable, and thus, they are no longer perfectly correlated. The bounds on the regret and Bayes risk and the policies found in [MER 09] no longer apply.

– *Bandits in metric spaces* [KLE 08, BUB 08]: we choose an arm $x_t \in X$ in metric space. The average reward $f(x_t)$ is assumed to be Lipschitz. It is comparable to $\sup_{x \in X} f(x)$. This is an optimization problem online. In [KLE 08], the authors considered generic metric spaces assuming that the mean-payoff function is Lipschitz with respect to the (known) metric of the space. They proposed an algorithm that essentially achieves the best possible regret with respect to these environments.

– *Hierarchical bandits*: algorithms UCT [KOC 06], BAST [COQ 07], HOO [BUB 08], application to the game of go [GEL 06].

Nowadays, there are lots of applications of MAB in different fields. Section 5.2.2 briefly analyzes some of them.

5.2.2. *Applications of MAB*

MAB problems play a key role for decision problems to maintain the "exploration versus exploitation" trade-off. The latter is considered as the gambling scenario described previously. After the gambler has known a slot machine to have good average payoff, there are two options to choose from: continuing to play this slot machine (exploitation) and trying other alternatives that have never been tested or that have only been tested infrequently (exploration). Since this type of trade-off appears widely in the field of online decision problems, there are many applications of MAB algorithms. Here we cite some applications, which we usually used, as follows:

– *Web search* [RAD 08]: for the Internet infrastructure, Web search has become an essential component and has attracted significant interest from the machine learning community. The conventional approach is associated with the learning-to-rank problem. In [RAD 08], an online learning approach for learning from usage data is presented. Since training data are being collected, it immediately impacts the rankings shown. The learning problem they address is regret minimization, where the goal is to minimize the total number of poor rankings displayed over all time. In particular, in this setting, there is a natural trade-off between exploration and exploitation. The authors propose a learning algorithm to calculate an ordering of search results from a diverse set of orderings. By iterating through all documents in each of the positions while keeping the documents in the other positions fixed, they attempt to learn a "best" ranking of documents using user clicks. Their approach naturally produces a diverse set of results, as user feedback through clicks will diminish the value of similar documents.

– *Design of ethical clinical trials* [HAR 92]: in this application, we have to evaluate n possible treatments for a disease in order to find which one is most effective. The appropriate solution is to design a clinical trial in assigning T subjects, partitioned randomly, into n groups of equal size. Each treatment is administered to one group of subjects, and the results across different groups are compared to see which treatment is most effective.

The experiment is launched over time so that some of the subjects arrive after the treatments have already been tested on earlier subjects. The rule is adjusted by assigning subjects into groups as time progresses. After some stages of the trial, a greater fraction of the subjects should be assigned to treatments that have performed well during the earlier stages of the trial. This increases the expected benefit of the treatments to the subjects in the clinical trial, and it also has the effect of increasing the amount of data collected for the most successful treatments. Because a subject's response to a treatment can always be evaluated before the next subject arrives, this problem becomes an MAB problem where the set of strategies is the set of treatments. The time periods are the sequence of subjects participating in the trial.

– *Internet ad placement* [GAB 09]: this is a case where a customer is launching an Internet advertising campaign for a new product. He has bought advertising space on a Website, one of n possible advertisements should be chosen to be displayed each time a user visits the site. If the user clicks on the advertisement, the payoff of displaying an advertisement to a user is 1, otherwise 0. In any case, this is an MAB problem in which the set of strategies is the set of advertisements that users might display.

– *Pricing*: a seller uses a Website to sell something with zero production cost. When a user visits the site, the seller offers a price and the user makes a purchase if and only if this price is less than or equal to the maximum amount the user is willing to pay. This is also an MAB problem in which the set of strategies is the set of possible prices. The payoff of the charging price is 0 if the user is unwilling to pay that price, or 1 if the user is willing to pay the price.

– *Overlay routing* [AWE 08]: in an overlay network, two hosts communicate in using a multi-hop path. The network experiences point-to-point delays that change unpredictably over time. Two hosts can only measure the end-to-end delay on a multi-hop path. The goal is to choose a routing path having a small delay time. That can be modeled as an MAB problem in which the set of strategies is the set of multi-hop paths. The cost of a path is equal to its end-to-end delay.

Since this book focuses on CDN service, the application we chose to apply the MAB is server selection method in a CDN system. In fact, there are many policies for applying the MAB to the applications mentioned above. Section 5.2.3 analyzes some widely used algorithms for the MAB problem. Then, we compare these algorithms to choose the best one, in terms of QoE, to apply to our CDN architecture.

5.2.3. *Algorithms for MAB*

MAB formalization of a sequential decision problem has to take into account the exploration versus exploitation dilemma, i.e. the balance between exploring the less chosen actions in order to find the most profitable actions while exploiting the currently identified best action as often as possible. Many algorithms provide policies to solve a MAB formalization of a given sequential decision problem. The measure that assesses the success of a policy is *the regret*. Lai *et al.* [LAI 85] indicate that the regret for MAB formalization has to grow at least logarithmically in the number of plays. In this section, we present three currently used algorithms for MAB formalization: *ϵ-greedy*, *softmax* and *upper confidence bounds*.

5.2.3.1. *ϵ-greedy algorithm*

First, the ϵ-greedy rule [WAT 89, KOU 08] (algorithm 5.1) is the simplest and most well-known policy for the bandit problem. ϵ-greedy policy includes randomly choosing a lever with ϵ-frequency, and otherwise choosing the lever having the highest estimated reward. More precisely, for this algorithm, the decision maker chooses, with a probability of $(1-\epsilon)$, the action with the highest average reward, and, with a probability of ϵ, a random action uniformly drawn in the set of those available.

Algorithm 5.1 ϵ-Greedy algorithm

Input: $[a_0, r_0, a_1, r_1, ..., a_{t-1}, r_{t-1}]$, $N \geq 1, 0 < \epsilon < 1$
Output: a_t
  Play each action once;
  for $i = 0$ to N **do**
    $\hat{r}_{a^i} \leftarrow$ current average reward of a^i
  end for
  if probability is $1 - \epsilon$ **then**
    return $a_t^i = \text{argmax}_i(\hat{r}_{a^i})$
  else
    return a_t^i is selected randomly
  end if

The simplest variant of the ϵ-greedy policy is called *ϵ-first policy*. The ϵ-first policy consists of realizing the exploration at the beginning. More precisely, for a number $T \in \mathbb{N}$ of rounds, one has a pure exploration phase where the actions are randomly chosen during ϵT first rounds. In the remaining $(1-\epsilon)T$ rounds, the action having the highest estimated average reward is chosen. We call it the pure exploitation phase. The ϵ must be in the interval $(0, 1)$ and it is chosen by the user. The PAC frameworks of [EVE 02] and [MAN 04] have presented the ϵ-first policy. There is another ϵ-greedy policy called *ϵ-decreasing policy*, which uses the decreasing ϵ for getting arbitrarily close to the optimal policy asymptotically. Like the general ϵ-greedy policy, a random action is chosen with ϵ frequency and the action having the highest average reward is chosen with the $(1 - \epsilon)$ frequency. The difference is the ϵ_t where t is the index of the current round. The decreasing ϵ_t is given by $\epsilon_t = \min\{1, \frac{\epsilon_0}{t}\}$ where $\epsilon_0 > 0$.

The value of ϵ_0 is chosen by the user. Cesa-Bianchi *et al.* [CES 98] use an algorithm called *GreedyMix* to analyze the ϵ-decreasing policy. The slight difference between the *GreedyMix* and ϵ-decreasing policy is the use of a decreasing factor of $\frac{\log(t)}{t}$ instead of $\frac{1}{t}$. The authors prove a $\mathcal{O}(\log(T)^2)$ regret for *GreedyMix* where T is the number of rounds. Auer *et al.* [AUE 02] improve this result and achieve a $\mathcal{O}(\log(T))$ regret for the ϵ-decreasing policy with some constraint of the choice of the value ϵ.

According to [AUE 02], ϵ_t is defined by

$$\epsilon_t = \min\left\{1, \frac{cK}{d^2t}\right\} \qquad [5.3]$$

where $c > 0$ and $0 < d < 1$ are parameters. For all $K > 1$ and for all reward distributions P_1, P_2, ..., P_K with support in $[0, 1]$, if policy ϵ-greedy is run with the input parameter: $0 < d \leq \min_{i:\mu_i<\mu_*} \Delta_i$, then the probability that after any number $n \geq cK/d$ of plays ϵ_t-greedy chooses a sub-optimal machine j which is at most:

$$\frac{c}{d^2n} + 2\left(\frac{c}{d^2}\ln\frac{(n-1)d^2e^{1/2}}{cK}\right)\left(\frac{cK}{(n-1)d^2e^{1/2}}\right)^{c/(5d^2)} + \frac{4e}{d^2}\left(\frac{cK}{(n-1)d^2e^{1/2}}\right)^{c/2} \qquad [5.4]$$

5.2.3.2. *Softmax algorithm*

Second, the Softmax policy [LUC 59, KOU 08] uses a function to assign average reward into action probabilities. The used function p^i is demonstrated in algorithm 5.2. The decision maker chooses an action based on probability p^i assigned to each action a^i. This algorithm allows us to explore less used actions at a given time but also focuses on best action in terms of gains expectation with respect to a Gibbs sampling distribution [GEO 11].

The action k is chosen with the probability:

$$p^i = \frac{e^{\frac{\hat{r}_{a^i}}{\tau}}}{\sum_{j=1}^{N} e^{\frac{\hat{r}_{a^j}}{\tau}}} \qquad [5.5]$$

where $\hat{r}$ is the estimated mean of the rewards brought by the action i and $\tau \in \mathbb{R}^+$ is a parameter called *temperature*. The user chooses the value of τ.

Algorithm 5.2 Softmax algorithm

Input: $[a_0, r_0, a_1, r_1, ..., a_{t-1}, r_{t-1}]$, $N \geq 1, 0 < \tau < 1$
Output: a_t
Play each action once;
for $i = 0$ to N **do**
$\hat{r}_{a^i} \leftarrow$ current average reward of a^i
$p^i = \frac{e^{\frac{\hat{r}_{a^i}}{\tau}}}{\sum_{j=1}^{N} e^{\frac{\hat{r}_{a^j}}{\tau}}}$
end for
return a_t^i is chosen with probability p^i

Like ϵ-greedy policy, the Softmax policy (Boltzman exploration) can be modified into *decreasing Softmax* whose temperature τ decreases with the number of rounds. More precisely, the decreasing Softmax has a temperature $\tau_t = \frac{\tau_0}{t}$ that depends on the index t of the current round. The user has to choose the value of τ_0. Bianchi *et al.* [CES 98] use the *Softmix algorithm* to analyze the decreasing Softmax. The Softmix differs from the decreasing Softmax with a $\frac{\log(t)}{t}$ factor instead of a $\frac{1}{t}$ factor.

There is another variant of the Softmax algorithm: the Exp3 "exponential weight algorithm for exploration and exploitation" [AUE 95]. At the round t, the probability of choosing the action k is defined as follows:

$$p_k(t) = (1 - \gamma)\frac{\omega_k(t)}{\sum_{j=1}^{K} \omega_j(t)} + \frac{\gamma}{K} \qquad [5.6]$$

where $\omega_j(t+1) = \omega_j(t) \exp\left(\gamma \frac{r_j(t)}{p_j(t)K}\right)$ if the action j has been chosen at time t with $r_j(t)$ being the observed reward, $\omega_j(t+1) = \omega_j(t)$ otherwise. The user chooses the parameter $\gamma \in (0, 1]$. The algorithm divides the reward $r_j(t)$ by the probability $p_j(t)$ that the action was chosen. In a modified version of *Exp3* [AUE 03] with decreasing γ. This approach achieved a regret of $\mathcal{O}(\sqrt{KT}\log(K))$.

5.2.3.3. *UCB algorithm*

Unlike the two algorithms mentioned above, the policy UCB1 (algorithm 5.3) [AUE 02] is related to an index-based policy. The UCB index is defined as the sum of the current average reward and the size of the one-sided confidence interval for the average reward within which the true expected reward falls with overwhelming probability. In the case where the reward generating process can be represented by a set of random variables, the confidence bounds provide an useful method to deal with the exploitation–exploration trade-off. The estimated means of the random variables represent the current knowledge and guide further exploitation. On the contrary, the widths of the confidence bounds reflect the uncertainty of the knowledge and then guide further exploration. Relating means and widths gives a criteria to decide when to explore and when to exploit.

Algorithm 5.3 UCB1 algorithm

Input: $[a_0, r_0, a_1, r_1, ..., a_{t-1}, r_{t-1}]$, $N \geq 1$
Output: a_t
  Play each action once;
  for $i = 0$ to N **do**
    $\hat{r}_{a^i} \leftarrow$ current average reward of a^i
    $UCBindex_{a^i} = \hat{r}_{a^i} + \sqrt{\frac{2\ln(N)}{n_{a^i}}}$
  end for
  return $a_t^i = \text{argmax}_i(UCBindex_{a^i})$

Lai *et al.* [LAI 85] prove that an optimal algorithm achieves a regret (equation [5.2]) $\rho(T) = \Theta(\log T)$ as $T \rightarrow \infty$ when the variances of the distributions Φ_i are finite. Agrawal [AGR 95] propose a simple learning algorithm to obtain such performance. The author used upper confidence bounds of the form $\hat{r_i}(t) + \sigma_i(t)$ for the expected rewards r_i of the distributions Φ_i. The $\hat{r_i}(t)$ is an estimate for the true expected reward r_i and $\sigma_i(t)$ is chosen such that $\hat{r_i}(t) - \sigma_i(t) \leq r_t \leq \hat{r_i}(t) + \sigma_i(t)$ with high probability. In each round t, the decision maker selects the action having the maximal upper confidence bound $\hat{r_i}(t) + \sigma_i(t)$. So, an action i is chosen if $\hat{r_i}(t)$ is large or if $\sigma_i(t)$ is large. If the $\sigma_i(t)$ is large, this choice is an exploration trial. On the contrary, if an action with large $\hat{r_i}(t)$ is selected, we call it an exploitation trial. The number of exploration trials is limited because the $\sigma_i(t)$ decreases rapidly with each choice of action i. If $\sigma_i(t)$ is small, then $\hat{r_i}(t)$ is close to r_i and an action is chosen in an exploitation trial only if it is the optimal action with maximal r_i. Therefore, the upper confidence bounds can really solve the exploitation–exploration trade-off.

We denote the average (or mean or expected) reward of the best action as μ^* and of any other action j as μ_j. There are a total of K actions. $T_j(n)$ is the number of times of trying action j in a total of n action. Formally, the regret after n actions is defined as:

$$\text{regret}(n) = \mu^* n - \sum_{j=1}^{K} \mathbb{E}[T_j(n)] \quad [5.7]$$

Regret is defined in terms of the average reward. So, if we can estimate average reward, then we can minimize regret. A confidence interval is a range of values within which we are sure that the mean lies with a certain probability. For example, we could believe that the mean is within $[0.2, 0.5]$ with probability 0.95. If we have tried an action less often, our estimated reward is less accurate so the confidence interval is larger. It shrinks as we get more information (i.e. try the action more often). Then, instead of trying the action with the highest mean we can try the action with the highest upper bound on its confidence interval. This is called an optimistic policy. We believe an action is as good as possible given the available evidence. There is one question: how do we calculate the confidence interval? We can turn to the classic Chernoff–Hoeffding bound [KAR 05] to get an answer. Let $X_1, X_2, ..., X_n$ be independent random variables in the range $[0, 1]$ with $\mathbb{E}[Xi] = \mu$. Then, for $a > 0$:

$$\mathbb{P}\left(\frac{1}{n}\sum_{i=1}^{n} X_i \geq \mu + a\right) \leq e^{-2a^2 n} \qquad [5.8]$$

The other side also holds:

$$\mathbb{P}\left(\frac{1}{n}\sum_{i=1}^{n} X_i \geq \mu - a\right) \leq e^{-2a^2 n} \qquad [5.9]$$

So, for $1 \leq n$ fixed:

$$\mathbb{P}\left(\hat{\mu_{k,n}} + \sqrt{\frac{2\log t}{n}} \leq \mu_k\right) \leq e^{-4\log(t)} = t^{-4} \qquad [5.10]$$

And also:

$$\mathbb{P}\left(\hat{\mu_{k,n}} - \sqrt{\frac{2\log t}{n}} \geq \mu_k\right) \leq e^{-4\log(t)} = t^{-4} \qquad [5.11]$$

The number of times that each action k is drawn, on average, at most is:

$$\mathbb{E}T_k(n) \leq 8\frac{\log n}{\Delta_k^2} + \frac{\pi^2}{3} \qquad [5.12]$$

So, the average cumulative regret of UCB is bounded as follows:

$$\mathbb{E}R_n = \sum_k \Delta_k \mathbb{E}T_k(n) \leq 8 \sum_{k:\Delta_k>0} \frac{\log n}{\Delta_k} + K\frac{\pi^2}{3} \qquad [5.13]$$

This result deduces that the average cumulative regret is logarithmic in n.

We have the uniform bound on the regret as follows:

$$\mathbb{E}R_n \leq \sqrt{8Kn(\log n + \frac{\pi^2}{3})} \qquad [5.14]$$

In this book, our server selection algorithm was inspired by the MAB problem. More precisely, we formalized the MAB to our server selection model and based it on UCB1 to design our selection algorithm. Besides the algorithm, we need a concrete scheme to implement our server selection method. Section 5.3 analyzes the provider-oriented scheme, which is a widely used scheme in today's CDN services. We will show their issues and explain why we have chosen the user-oriented scheme for our proposal.

5.3. Server selection schemes

Nowadays, there are lots of CDN providers with different strengths and weaknesses all based on many different needs of a specific customer. Some representative providers include:

– Akamai [AKA 98]: who offer distribution services such as http content and streaming media as well as other services such as network monitoring and geographic targeting. Akamai purchased InterVu (April 2000), Speedera (in 2005) and Netli (in 2007). These are some big customer names of Akamai: Adobe, Yahoo, IBM, NBA, Travelocity, Sony, CNET, NTT, Myspace, US Airforce, Monster, AMD, Reuters, NASA, etc.

– Limelight Networks [LIM 01]: this provider gives an advanced CDN for the Internet distribution of high-bandwidth media such as video, music and games. It also has some customers such as Amazon, Akimbo, iFilm, MSNBC, LonelyPlanet, Valve and VideoJug.

– EdgeCast [EDG 06], this CDN provider offers video, games, music, content as well as live events solutions. It also serves Website acceleration to increase Website performance and decrease page loading times. Its customers are some Web 2.0 companies such as Jaxtr, Mashable, Blip.tv and Mahalo.

Besides different implementations, the majority of CDN providers actually apply similar DNS-based server selection schemes [PAN 03]. These schemes are transparent to end users. This kind of server selection is based on the DNS resolution[1].

1 Domain Name System (DNS), defined as a hierarchical distributed naming system for computer networks, is an essential and effective component of the functionality of the Internet. It uses a domain name service to change queries for these names into IP addresses for the purpose of locating computer services and devices worldwide [WIK 01].

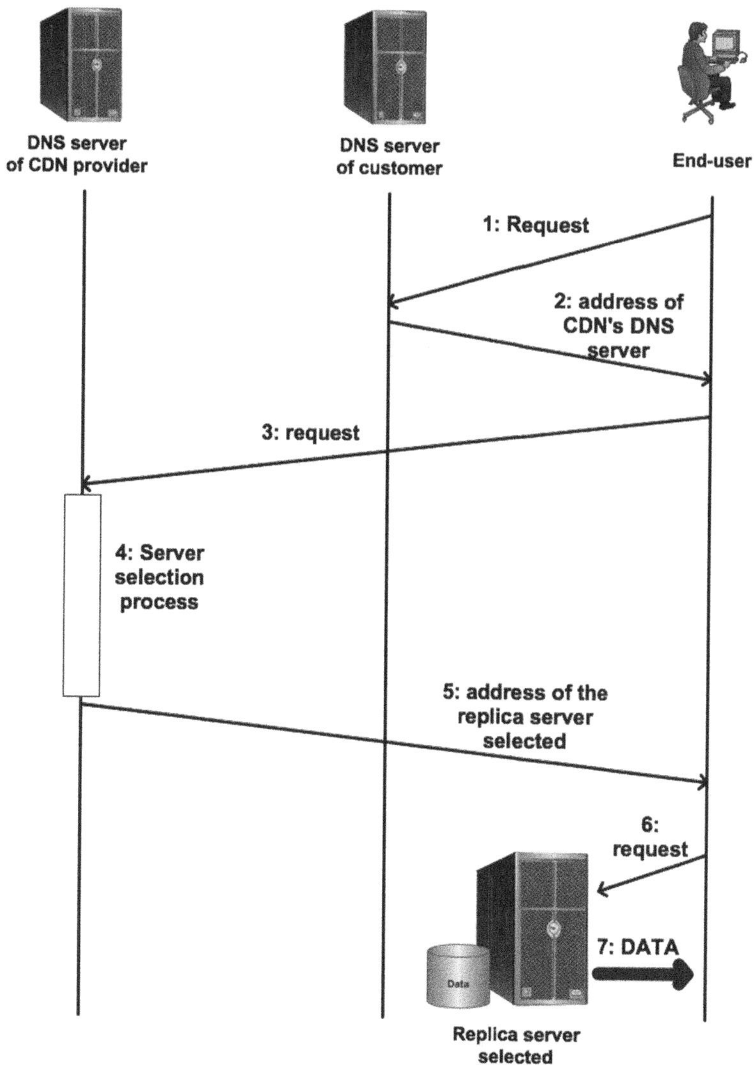

Figure 5.1. *DNS-based server selection scheme*

The DNS-based server selection scheme has three principal characters: *the end user*, *DNS server of customer* and *DNS server of CDN provider*. As shown in Figure 5.1, this scheme is described step-by-step as follows:

– Step 1: The user sends his/her request to the DNS server of the customer after some basic DNS resolution phases.

– Step 2: The DNS server of the customer replies to the user with the DNS server's address of the CDN provider.

– Step 3: After receiving the address, the end user sends the request to the DNS server of the CDN provider.

– Step 4: The server selection process is launched to choose an appropriate server for the user.

– Step 5: The CDN provider sends the address of the server selected to the end user.

– Step 6: The user sends the request to the replica server selected.

– Step 7: The replica server provides data to the end user.

Since the server selection is launched at the DNS server of the CDN provider, this approach is considered as *provider-oriented* server selection. In other words, the CDN provider makes a decision of replica server to serve the end user. In [PAN 03], this kind of server selection advantageously uses the existing DNS infrastructure by making no change at either the context provider or end user side. That makes this approach immediately deployable in the Internet. Despite this, provider-oriented server selection has the following disadvantages:

– This server selection scheme operates at the application layer. So, the IP addresses of selected replica servers are treated differently. Definitively, with certain applications that are not interested in the Time To Live (TTL), an IP address of the replica server is used deterministically, regardless of how the CDN provider dynamically switches between selected servers. In other words, the end user is always served by the replica server selected even when network condition and server load change dramatically during the session. This fact reduces the effect of dynamic selection of the DNS-based method.

– There is a problem coming from the TTL setting of DNS record. If the TTL is small, the end user must send more requests to the DNS server. That causes more traffic and more DNS-related delay. Consequently, this approach faces the congestion and packet loss issues.

– Requests coming from the same local DNS of different clients are treated equally. The DNS does not take into consideration the protocol and object identifiers in requested uniform resource locators (URLs). As a result, there is no distinction between different services such as http and ftp, even if each service has proper requirement.

– For the DNS-based approach, the CDN provider manages users as clusters. Each cluster is managed by a local DNS server, which takes into account the DNS-related resolution phase. There is a critical assumption of DNS-based server selections: all users are close to their local DNS server. Then, all users managed by a local DNS server are treated equally. However, this assumption is not always true, especially in the actual Internet with large corporations and ISP networks having wide coverage, multiple connections. This fact is proved in [MAO 02] who found that only 64% of all users have a local DNS server with the same autonomous system (AS) number. In [SHA 01], the authors prove that the user-DNS proximity assumption is far from realistic in showing that 15% of Web clients and its DNS server pairs are in 1-hop clusters. They also showed the median cluster size is 5 hops.

– Since DNS servers cannot identify the location of local DNS servers and do not have complete knowledge about the network status and connectivity, provider-oriented server selection cannot select the most appropriate replica server, even when end users are close to their local DNS server.

The disadvantages mentioned above motivate us to apply a user-oriented server selection scheme to the CDN system. According to Yoshikawa *et al.* [YOS 97], "in many cases the client, rather than the server, is the right place to implement transparent access to network services". Our proposed scheme is based on QoE feedback of end users and selects the replica server on the user side. Theoretically, it will resolve all these above-mentioned problems.

5.4. Our proposal for QoE-based server selection method

Server selection is an important operation in any CDN, aiming to choose the appropriate server according to some predefined metrics. Indeed, if an inappropriate server was chosen (e.g. the overloaded server or the very far one), the quality of delivered data would quick degrade. Previous research on server selection methods are mainly based on QoS metrics such as RTT, hop-count between a client-server pair and loss rate. However, these studies did not take into account a crucial concept that constitutes a determining factor for success of the deployment of future services, which is based on user perception: the QoE.

In sections 5.4.1 and 5.4.2, we present the proposed server selection scheme and our UCB1-based selection algorithm.

5.4.1. *Proposed server selection scheme*

In this section, we present our QoE-based CDN model (Figure 5.2), which operates like other CDN models, except that it is based on QoE feedback of end users instead of only QoS parameters.

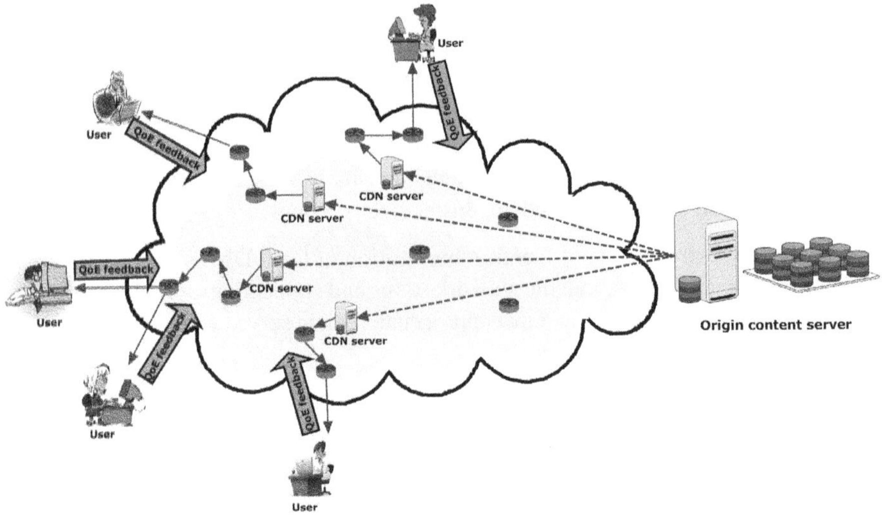

Figure 5.2. *QoE-based CDN model*

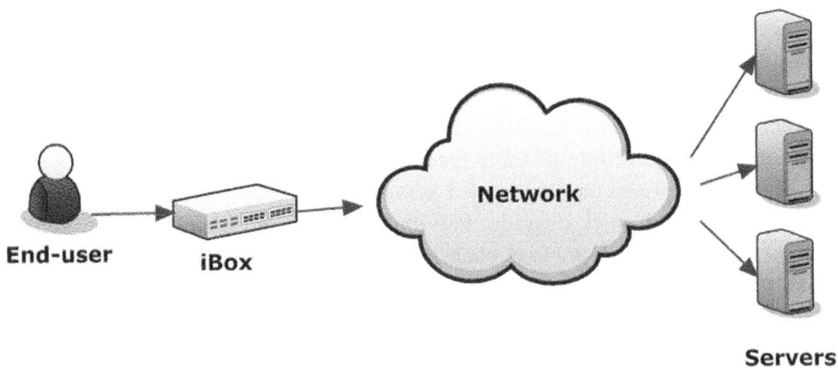

Figure 5.3. *CDN model with four components*

The model (Figure 5.3) is organized into four components: *end user*, *iBox*, *network* and *servers*. The iBoxes are placed on the user side, typically serving only one user. The users send requests to the system. These requests are transparently passed through the associated iBox. Based on information in its database and on a server selection policy, the iBox realizes the server selection process and chooses the server to forward the request. The iBox also supports the QoE measurement method. Each iBox has a QoE table that consists of MOS values, a composite metric representing QoE score. Each MOS entry is corresponding to a replica server. So, the number of MOS entries is equal to the number of replica servers.

To apply our proposed approach to another CDN architecture, note that the iBox we used in our model is not necessarily a piece of network equipment. That may be a piece of software embedded in any piece of equipment with the same functionality.

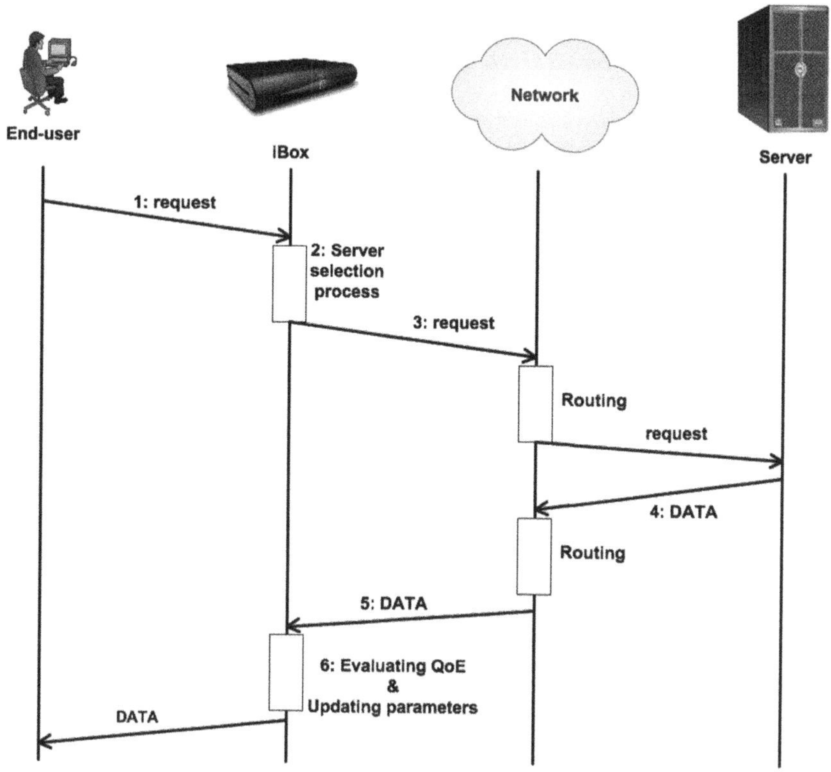

Figure 5.4. *QoE-based server selection scheme*

We present a step-by-step description of the server selection operations (Figure 5.4):

1) The user sends a request to his/her iBox.

2) This iBox uses the server selection algorithm to choose the adequate server based on its MOS table. In this step, we implement several algorithms of server selection, based on the state of the art, to compare their performances.

3) After assigning the selected server in the request packet, the iBox forwards it to the network.

4) The chosen server receives a data request and serves the requested data to the user. Without loss of generality, we assume that all the requested data can be found in any replica server.

5) The network system forwards data from the server to the associated iBox.

6) After receiving data from the server, the iBox evaluates the QoE to update parameters if this server is in the selection table.

5.4.2. *Proposed UCB1-based server selection algorithm*

We decide to formalize the server selection method into the MAB formalization presented previously.

We now present the numerical details of our server selection based on algorithm UCB1 (algorithm 5.3). As explained previously, our selection policy aims at selecting servers based on indexes that provide UCB on the rewards (MOS score) associated with the server.

Each iBox has an MOS table that contains MOS scores of the servers. The iBox uses this table to select server in using the UCB1 algorithm. In our model, we have N servers, s_i with $i = 0...(N-1)$. In the time t, each iBox has a vector of information $\{s_0, M_0, s_1, M_1, ..., s_{t-1}, M_{t-1}\}$ where s_j and M_j are, respectively, selected servers and the measured MOS score in time j . The general approach to assess the average reward (MOS score) of a server s_i is to consider a confidence bound for its sample mean. Let $\hat{M_i}$ be the sample mean of server s_i after being selected n_{s^i} times. The steps to select server (applied for all of iBoxes) are given below:

1) Each server is selected at least once.

2) Calculate the current average MOS score of each server

$$\hat{M_i} = \frac{\sum_{j=0}^{t-1} M_j \times \mathbf{1}_{\{s^i \text{is chosen}\}}}{n_{s^i}} \qquad [5.15]$$

where $\mathbf{1}_{\{s^j \text{is chosen}\}}$ is equal to 1 if the server s^j is chosen, 0 otherwise. n_{s^i} is the number of chosen times of the server s_i.

3) Calculate the upper bound confidence index, UCB_i, for each server s^i:

$$UCB_i = \hat{M_i} + \sqrt{\frac{2\ln(t)}{n_{s^i}}} \qquad [5.16]$$

For all servers, UCB_i gives an optimistic estimation of the expected reward obtained when the iBox selects the server s_i at a time t after being tested n_{s^i} times.

4) Choose the server having the maximum UCB index: $s_t = \text{argmax}_i(UCB_i)$.

As shown in formula [5.16], the UCB index is composed of two parts: the average MOS score of server i, $\hat{M_i}$, and the confidence bound $\sqrt{\frac{2\ln(t)}{n_{s^i}}}$. Hence, the server i

is chosen if $\hat{M}_i$ is large or if $\sqrt{\frac{2\ln(t)}{n_{si}}}$ is large. If the average MOS score is large, this choice is an exploitation trial. On the contrary, if a server with a large confidence bound is selected, it is an exploration trial. The confidence bound is large if the number of chosen time of server i, n_{si}, is small compared to the total chosen time t. In other words, the less the server is chosen, the more it has opportunity to be chosen.

5.5. Experimental results

To validate the proposal, we used the same methodology as routing experiments. More precisely, first, we tested the selection algorithm in a simulation model to validate the functionality of our proposal. Then, we began implementing a real platform to verify its performance and compare it with other approaches. Sections 5.5.1 and 5.5.2 present the obtained experimental results.

5.5.1. *Simulation results*

We first analyze the obtained simulation results. We compare our approach with the following three server selection policies that have been applied in previous research efforts. As explained later, these approaches are mainly based on the optimization of QoS performance metrics:

– *FASTEST policy* [WEN 10, HAN 09]: the iBox always chooses the server that has the shortest RTT (i.e. the time it takes for a data packet to be sent and an acknowledgment of that data packet to be received) between a client and a server. The RTT value between client-server pair is updated online.

– *CLOSEST policy* [PIE 06, GUO 02]: the closest server of the iBox in terms of hop-count (i.e. the number of routers through which data must pass between user and server) is always chosen. The hop-count between the client-server pair is updated online.

– *ROUND ROBIN policy* [RIN 09]: each iBox has a server list. For a round robin policy, the first request is sent to a server picked randomly from the list. For subsequent requests, the iBox follows a circular order fixed before selecting the server.

The reason we chose these server selection policies mentioned above is that they are widely used server selection algorithms of today's CDN providers. For example, the FASTEST policy is used by Akamai [AKA 98] and NLARN [NLA 06] which is developing a project of server selection based on RTT. The CLOSEST policy is implemented in project Globule [PIE 06] that is based on the hop-count for the selection of the appropriate server. The ROUND ROBIN policy is applied to the Ring Server project.

In this chapter, we have presented three currently used algorithms for MAB formalization: ϵ-*greedy*, *softmax* and *UCB*1. We now compare their performances to choose the best one to apply to our proposal.

The experiments are launched in the platform presented in Chapter 4 in section 4.7.1 by using two network topologies: the *3AreasNet* and the *NTTnet* (Figure 4.19). Figures 5.5a and b show that UCB1 is explicitly better than these two policies in term of MOS score. The UCB1 reaches mean MOS value of 4.0 in Figure 5.5a and 3.7 in Figure 5.5b. After the fifth hour, UCB1 becomes increasingly better than the other two. UCB1 gives a MOS score that varies between 3.9–4.0 (Figure 5.5a) and 3.7–3.8 (Figure 5.5b). The average MOS score of UCB1 is always higher than a score of 3.5.

However, the ϵ-greedy gives the worst result with the MOS score between 2.4–2.5 in the two figures (Figures 5.5a and b). The Softmax obtains the MOS score of 3.3 and 3.1 in Figures 5.5a and b, respectively.

Obviously, the above results show that the algorithm UCB1 outperformed the two other algorithms in terms of MOS score. That motivates us to apply the UCB1 to our server selection method.

In Figure 5.6, the UCB1 algorithm has a mean MOS value that is greater than 3.5, while the other three have a value less than 3.5. For example, in the 28th hour of the simulation on the *3AreasNet* (Figure 5.6a), the obtained MOS scores are 3.9, 3.3, 2.7 and 2.4 for UCB1, FASTEST, CLOSEST and round robin, respectively (Tables 5.1 and 5.2).

	UCB1	CLOSEST	FASTEST	ROUND ROBIN
Average MOS	3.9	2.7	3.3	2.4
Convergence time (s)	210	80	100	–
Improvement of UCB1 (MOS)		144.44%	118.18%	162.5%

Table 5.1. *Experimental results for 3AreasNet*

	UCB1	CLOSEST	FASTEST	ROUND ROBIN
Average MOS	3.7	2.4	3.0	2.05
Convergence time(s)	220	90	120	–
Improvement of UCB1 (MOS)		154.16%	123.33%	180.48%

Table 5.2. *Experimental results for NTTnet*

According to the round robin policy, as shown in Figures 5.6a and b, round robin gives the worst result (MOS score of 2.4 and 2.05), which results in the selection of inappropriate servers. For example, all servers are chosen periodically even if a server has a long routing path to the user or it is already overloaded.

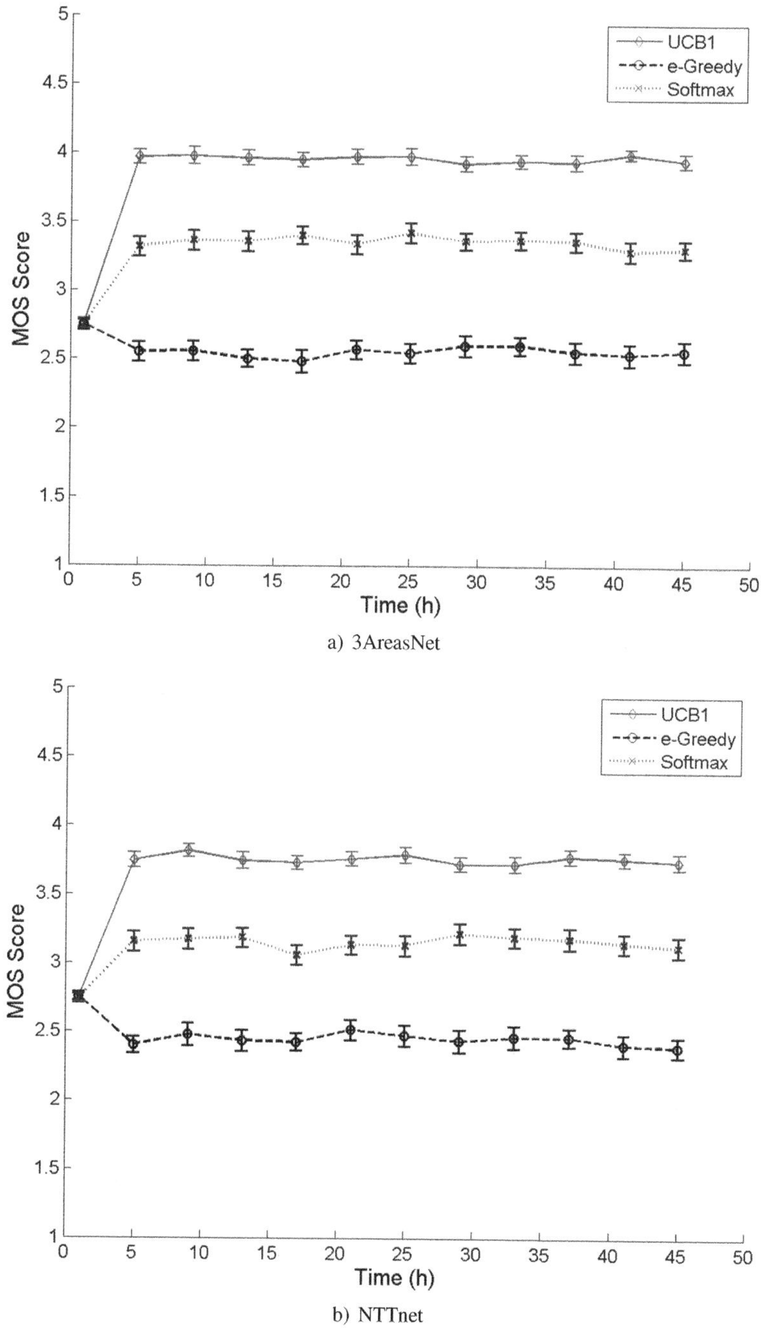

Figure 5.5. *User perception results of three policies: ϵ-greedy, softmax and UCB1*

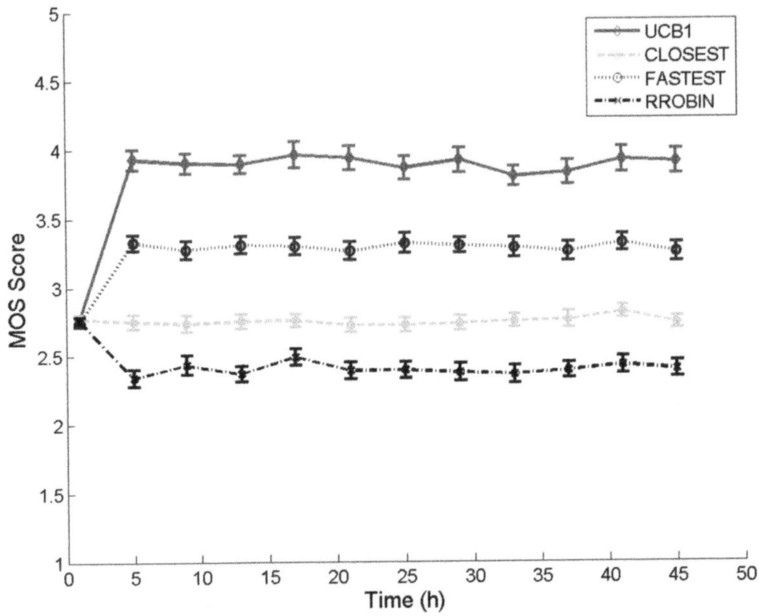

a) 3AreasNet

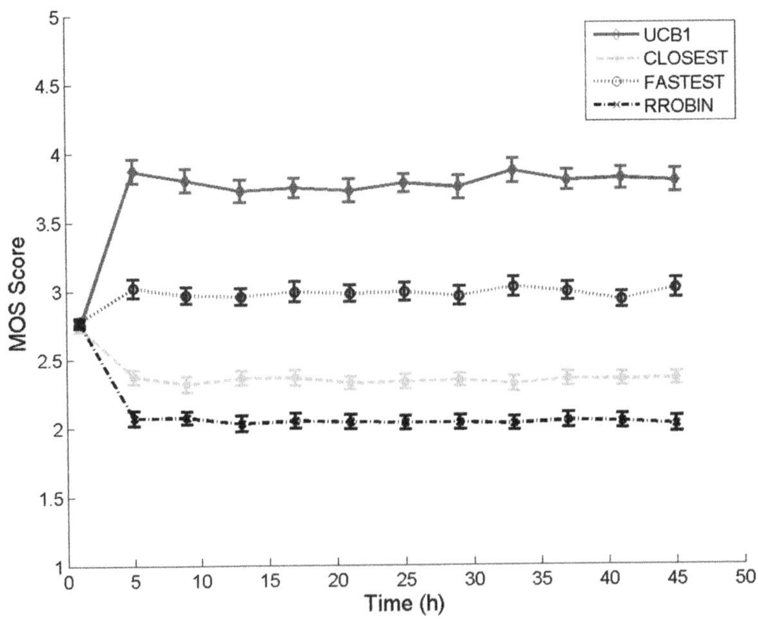

b) NTTnet

Figure 5.6. *User perception results of four policies: UCB1, FASTEST, CLOSEST and round robin*

We now analyze the results of the CLOSEST algorithm that always selects the closest server in terms of hop-count. In many cases, the closest server does not give good results in terms of service quality (e.g. in the case where a server is close to many users). Taking an example in the 3AreasNet topology (Figure 4.19a), the server *sv47* is close to four users *u53*, *u40*, *u60* and *u54*. In contrast, in the NTT topology as shown in Figure 4.19c, the server *sv78* is close to five users *u56*, *u57*, *u58*, *u59* and *u60*. For the CLOSEST algorithm, all the users select the closest server. This fact overloads the selected server and leads to the degradation of quality. This explains the weak result obtained from the CLOSEST algorithm in Figure 5.6. The decrease in MOS score from 2.7 (Figure 5.6a) down to 2.4 (Figure 5.6b) is explained by the change of topology. The NTTnet topology is based on a real network called NTT where choosing the closest server is not always suitable (in the case of server *sv78*).

The FASTEST algorithm gives a quite good result (3.3 in Figure 5.6a and 3.0 in Figure 5.6b). This algorithm always selects the server having the shortest RTT. As mentioned previously, there are four criteria that influence the service quality: *delay time*, *jitter*, *loss rate* and *conditional loss rate*. Taking into account only one of these three criteria is not enough to give a good QoE.

Results in Figures 5.6a and b show that the UCB1 algorithm obtains good QoE feedback with the MOS score greater than 3.5 (it is worth noting that the MOS score of 3.0 corresponds to fair quality).

Concerning the convergence time issue, in the first 5 min of simulation, we observe MOS score results of four algorithms every 10 s for two topologies (Figure 5.7). Note that the time unit used in Figures 5.7a and b is in “s”. We observe that four policies have a common initialization period where the policies do not have any information about the system. We call this period the *initialization phase*. As each algorithm has a particular policy, the convergence time of each algorithm is different.

For the 3AreasNet network (Figure 5.7a), after approximately 210 s, the UCB1 is stable with a MOS score of 3.9. The convergence times of FASTEST and CLOSEST are 100 s and 80 s, respectively. To explain the strong fluctuation in this period, we observed the UCB1 in the first 220 s (Figure 5.8). The Figure 5.8 shows the number of choices of one iBox in our experiment (iBox 55) for the first 220 s in the 3AreasNet topology. In this case (Figure 4.19a), the minimum hop-counts from iBox55 to the five servers *sv45*, *sv44*, *sv10*, *sv47* and *sv46* are 4, 5, 6, 6 and 7, respectively. In the first 70 s, *iBox55* has mainly chosen the *server45* due to the fact that *server45* is the closest server. Then, from the 100th second, *server10* is the most chosen server. The overloading of *server45* caused by choices of *iBox56*, *iBox51* and *iBox48* explains this selection switch. In addition, the number of choices of *server47* and *server46* decreases because the data flows have passed through the link *8-13*. The latter is highly overloaded because it is the connected link between two zones. This kind of adaptive behavior possibly explains the results obtained from our proposed method.

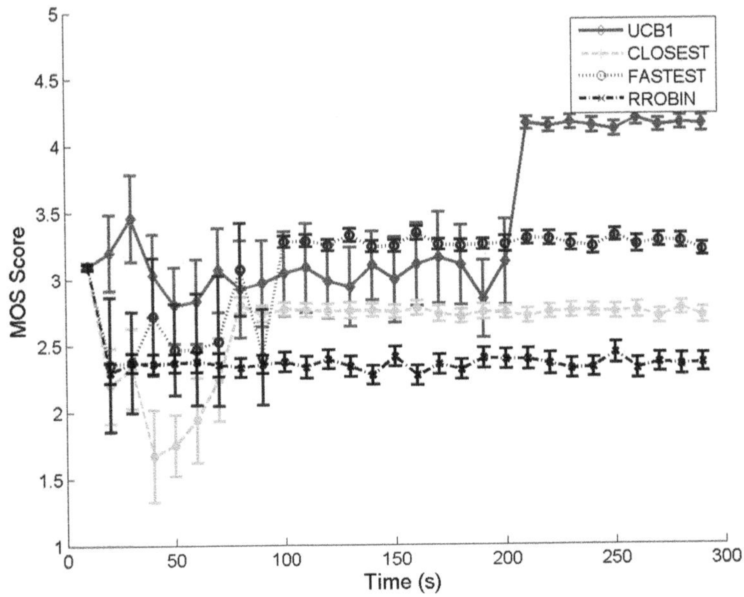

a) 3AreasNet

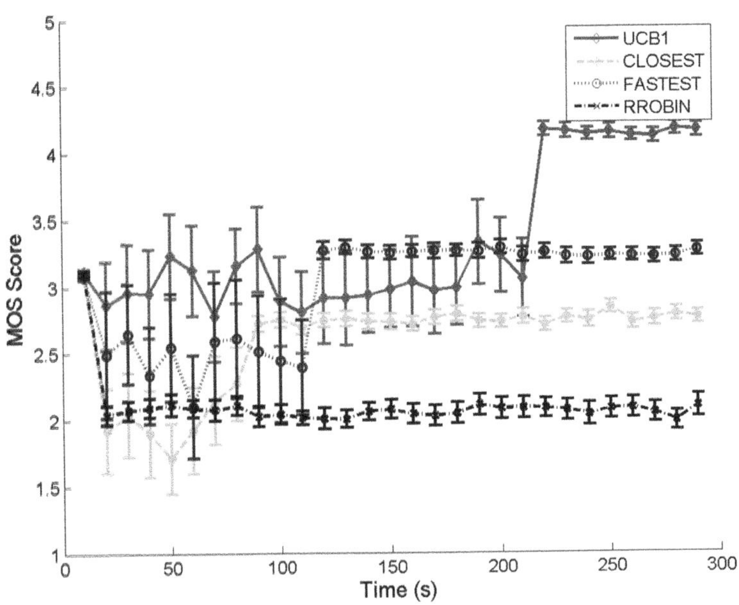

b) NTTnet

Figure 5.7. *Initialization phase*

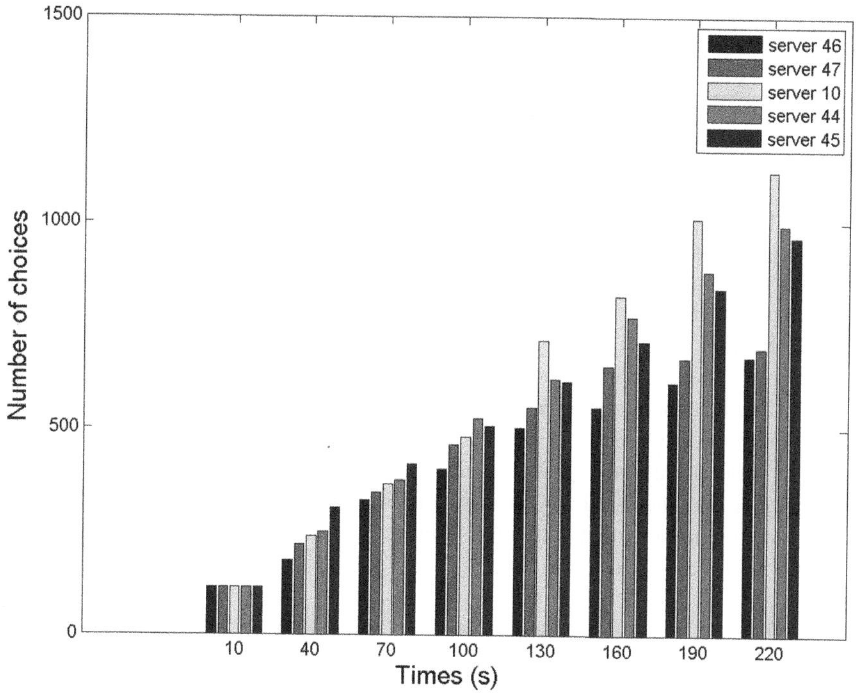

Figure 5.8. *Number of choices of iBox 55*

For the NTTnet network (Figure 5.7b), UCB1 reaches the MOS score of 3.7 after 220 s. The convergence time of FASTEST algorithm is 120 s and its MOS score is 3.0. The CLOSEST algorithm takes 90 s respectively, during the initialization phase.

5.5.2. *Real platform results*

We now analyze the results obtained of the real platform. We used a video of 4,000 s. The server selection module is executed every 20 s. So, we divided the video into 200 tracks, each one lasting 20 s. To evaluate the robustness of the proposed approach when the network environment changes, we made a change in the network context at about the 2000th second for each test (we marked a vertical line in Figures 5.9a and b). Making a change in the network context is achieved by changing the network parameters for each server. Here we give an example in Table 5.3 that shows the network parameters for 11 servers (from *sv1* to *sv11*) in two contexts. The three numbers in each box are *jitter*, *loss rate* and *conditional loss rate*. As shown in Table 5.3, the parameters are changed in passing from *Context 1* to *Context 2*. Figure 5.9 shows the results of two executions of the test.

Context	sv1	sv2	sv3	sv4	sv5	sv6	sv7	sv8	sv9	sv10	sv11
1	6 3 18	10 3 20	4 5 10	15 2 50	20 5 10	12 4 25	9 5 16	7 4 32	8 3 27	7 4 28	6 3 31
2	4 3 15	10 1 20	8 4 25	5 3 18	0 2 10	7 3 15	6 5 24	7 4 18	6 4 32	7 5 35	2 4 19

Table 5.3. *Example of change of network parameters*

In Figure 5.9a, the UCB1 algorithm takes about 200 s, from 0th to 200th second, to converge to the MOS score of 3.7. In Figure 5.9b, this initialization phase is from 0th to 220th second and the converged MOS value is 2.91. We can observe that during this first period (marked "Context 1" in the two figures), the values 3.7 and 2.91 are the best obtained values (among 11 servers) in Figures 5.9a and b, respectively. Due to the exploration task, UCB1 sometimes chooses another server at 740th, 880th, 900th, 1060th, 1280th second, etc., in Figure 5.9a and at 300th, 340th, 380th, 420th second, etc. In other words, UCB1 does not always choose the server giving the best result to guarantee the exploration–exploitation trade-off. The frequency of switching in Figure 5.9b is higher than the frequency of switching in Figure 5.9a. This is explained by the fact that the maximum value obtained in Figure 5.9a is greater than the value obtained in Figure 5.9b (i.e. 3.7 > 2.91).

In the case of the context change (the vertical line with the text "context change"), the UCB1 algorithm takes 320 s (from 1920th to 2240th second) as shown in Figure 5.9a and 240 s (from 1760th to 2000th second) as shown in Figure 5.9b for the adaptation phase. In the second context (marked "Context 2" in two figures), the converged MOS value is 4.34 and 3.71 as shown in Figures 5.9a and b, respectively. As shown in both figures, there is no switching task from 2240th to 3820th second (Figure 5.9a) and from 2000th to 3780th second (Figure 5.9b) because these two values (3.71 and 4.34) are already good in the scale of MOS score of 1–5 for video quality at the client side.

For the round robin and the FASTEST algorithms, the round robin algorithm follows a round robin rule in choosing the servers. However, the FASTEST algorithm chooses the server that gives the shortest delay. We can see that the fastest one is not the best one because the video quality does not depend only on the delay parameter.

These experimental results demonstrate that our proposed selection policy based on MAB formalization explicitly optimizes the QoE result. The results obtained also show that the proposed adaptive control model I.1 gives a better QoE result (the MOS score) than traditional models based on QoS metrics only.

5.6. Acknowledgment

The authors would like to thank Dr. Julien Perez for his fruitful help in conducting the work done in this chapter, especially for MAB formalization.

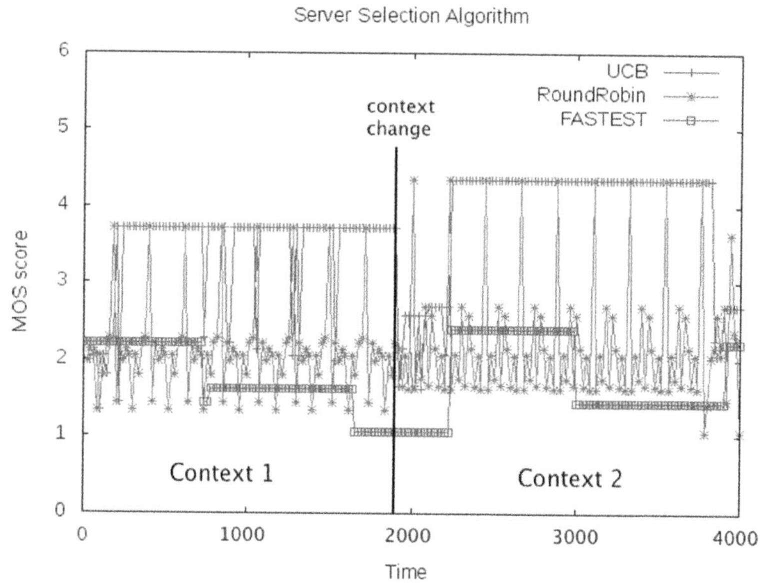

a)

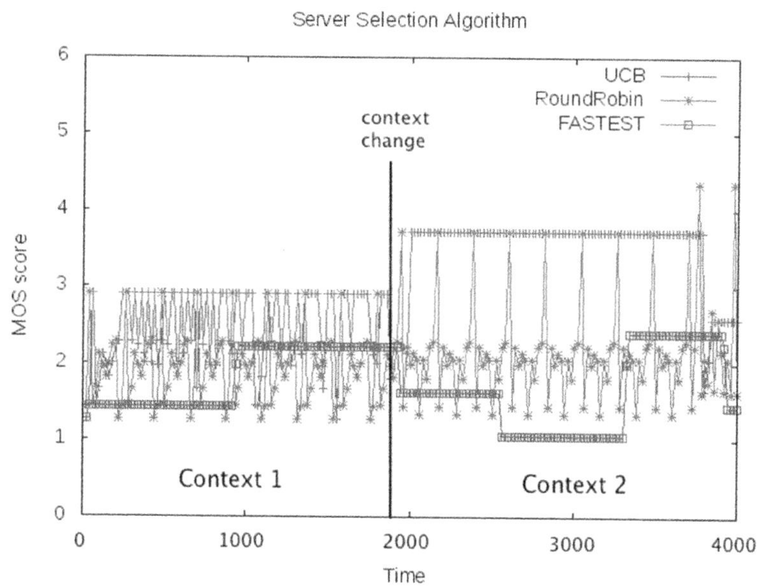

b)

Figure 5.9. *Experimental results over the real platform*

5.7. Conclusion

In this chapter, we focused on an example of a server selection algorithm based on the QoE notion, especially applied to the meta-routing layer of a CDN. The approach is implemented in QoE-based CDN architecture (QCDNA). The machine learning MAB formalization allows us to explicitly maximize a QoE measurement without any assumption. As shown in the experiment results, the algorithm based on QoE gives the best result in terms of user perception. It is also an optimal decision method in terms of minimizing the regret.

Conclusion

Network services have grown increasingly over the past several years with a variety of high-quality applications such as IPTV, VoIP, videoconferencing, telepresence and online games. Facing such rapid development, the widely used Quality of Service (QoS) framework, which is defined as an aspect ensuring service quality of network traffic, is not suitable enough. In fact, the traditional QoS networking is based on the technical parameters (e.g. delay, jitter, bandwidth and loss rate), which are too trivial to be competitive factors between service providers in today's Internet. Actually, a traffic flow delivered with excellent QoS metrics may risk delay experience of a videoconference client or an echo of the voice of a VoIP user. In addition, satisfying such a set of constraints risks meeting problems of complexity due to the fact that a QoS model with more than two non-correlate criteria meets an NP-complete problem. All service operators and providers gradually know that the most important goal is to satisfy end users who directly "pay" for these services. Ignoring user perception risks client dissatisfaction, churn and lost revenue. Therefore, satisfying end users today becomes a new trend for network management systems. Service providers need a new criterion that refers to the user experience in order to improve the traditional QoS parameters. In this book, we showed the inefficiency of QoS-based algorithms for user satisfaction optimization. As the latter is, by definition, directly related to end users, it seems to be reasonable to describe it as a nonlinear function of the QoS parameters. In addition, the QoS is important but not determinant in order to realize the mechanisms, allowing us to obtain and optimize the user satisfaction that is the important goal of operators. Consequently, Quality of Experience (QoE) has been introduced to meet this request. This new notion is defined as levels of user satisfaction when using the network services. Unlike the QoS networking, which focuses only on technical parameters, the QoE networking navigates the target to end users in ensuring an acceptable experience level. More precisely, the Internet today, mostly the content-oriented network (CON), needs an efficient end-to-end QoE model to assure the satisfaction of end users. An end-to-end QoE model is able to guarantee the user satisfaction between network hosts. Proactive monitoring enables the system to detect, in real time, the degradation of the QoE and to react adaptively to avoid the system's fall.

Conversely, over the years, the continuous technological evolution and the development of new applications and services have steered networking research towards new problems, which have emerged as the network evolves with new features towards what is usually referred to as the Future Internet, which has become one of the basic infrastructures that supports the world economy nowadays. In fact, there is a strong need to build a new network scenario, where networked computer devices are proliferating rapidly, supporting new types of services, usages and applications: from wireless sensor networks and new optical network technologies to cloud computing, high-end mobile devices supporting high-definition media, high-performance computers, peer-to-peer networks and various platforms and applications. This new network scenario is fueling research in the area of new network architectures that consider both the requirements and demands of key emerging applications and services, such as cloud computing and Internet video, and the currently deployed network infrastructures.

Therefore, the overall challenge is to find scalable and sustainable solutions for the ever-growing smart communications field, which support different kinds of services, for a wide variety of future next-generation network applications and cloud computing technologies. To address this challenge, new and cross-disciplinary approaches are required to optimize the distribution of media data. One approach is to introduce and combine methods for more effective storage of the data close to the consumer and various types of peer-supported distribution mechanisms. Decisions on when and where data are temporarily stored should be based on both estimates of expected demand and the final consumer demands on perceived quality. Also, statistical methods for online estimation of consumer demands will be crucial. Many of these issues are complex and can, in the short term, hardly be solved by any single approach. We need to find solutions to deliver network services in the most efficient way to provide users with the best perception while taking into consideration scarce network resources.

Our vision of a new paradigm is to make interactions first-class objects from the perspective of the user, the application and the network components. This is achieved by analyzing the interaction between the user and the application with quality perception metrics that are used to fix the control/command chain in network components. The idea here is how to integrate these metrics into a control/command chain in order to construct a network system. The key idea is to design an adaptive loop for improving the service quality of a network system by taking into account the end user feedback. In this model, the perception measurement module evaluates the user feedback and sends the result to the control module. The latter analyzes the received results, makes a decision and sends it to the command module that initiates the chosen action on the network system. This loop model refers to an adaptive system based on the end-to-end user perception and results in smart communications between equipment.

To show the validity of this approach and to meet the requirement mentioned above, this book focuses on such a model based on an end-to-end QoE model and applied to a content distribution network (CDN) architecture. The CDN is an effective approach for improving the service quality of the Internet by replicating the content from the place of origin to the replica servers scattered over the network system. So, the user's requests are served by replica servers close to the user's location. Despite this, a CDN service should guarantee an end-to-end QoE to satisfy the end user who plays a key role in the network management. The end-to-end proposed model is based on QoE with an adaptive control including three modules: QoE measurement, control and command. The QoE measurement evaluates the QoE based on the user feedback. In its turn, the control module takes the QoE value evaluated to make a decision by choosing an appropriate action. Finally, the command module applies this chosen action to the environment. This adaptive loop makes the system become an evolutionary system, which operates adaptively and infinitely in order to maintain a sufficient QoE level of the entire system. The model is also able to take into account the user's perception and network changes. The goal of the adaptive control is to optimize the QoE level on the user side.

User-driven network control aims to produce a new paradigm of smart communications based on multiservice user perception estimator agents. The estimators will be based on models of human perception and upon service discovery, trained progressively from the best-fitting current profile for that service. Each individual agent will communicate with other user agents through the network, exchanging profile information, so that services and applications could adapt to both end preferences and expectations on a peer-to-peer basis. Also, by participating in the agent's network, the service provider can obtain relevant profile user information (by groups or individuals) so as to be able to adapt resource provisioning from the network and thus optimize both the user experience and the cost of operation (e.g. in a cloud paradigm: the amount of computing resources, the bandwidth allocated to a particular service for a user or a group of users, the physical resource location, etc.). By having a good estimation of user perception, the provider can also better define the objectives of the service being provided, and translate them into the more technical objectives of the service components that need to be provisioned. In addition, evolved service level agreements (SLAs) can be defined in terms of this user's perception so that both the service provider and the user have a better way of following up and measuring if the objectives are being met.

Bibliography

[AGB 08] AGBOMA F., LIOTTA A., "QoE-aware QoS management", *Proceedings of the 6th International Conference on Advances in Mobile Computing and Multimedia (MoMM '08)*, Linz, Austria, 2008.

[AGR 95] AGRAWAL R., "Sample mean based index policies with O (log n) regret for the multi-armed bandit problem", *Advances in Applied Probability*, vol. 27, no. 4, pp. 1054–1078, 1995.

[AKA 98] AKAMAI, Akamai, available at http://www.akamai.com, 1998.

[ALM 04] ALMEIDA J., EAGER D., VERNON M., *et al.*, "Minimizing delivery cost in scalable streaming content distribution systems", *IEEE Transactions on Multimedia*, vol. 6, no. 2, pp. 356–365, 2004.

[AND 02] ANDREWS M., SHEPHERD B., SRINIVASAN A., *et al.*, "Clustering and server selection using passive monitoring", *Proceedings of the 21st IEEE International Conference on Computer Communications (INFOCOM 2002)*, New York, 2002.

[ANT 08] ANTOS A., GROVER V., SZEPESVÁRI C., "Active learning in multi-armed bandits", *19th International Conference on Algorithmic Learning Theory*, Springer, Budapest, Hungary, 2008.

[ARD 01] ARDAIZ O., FREITAG F., NAVARRO L., "Improving the service time of web clients using server redirection", *ACM SIGMETRICS Performance Evaluation Review*, vol. 29, no. 2, pp. 39–44, 2001.

[ARO 12] AROUSSI S., BOUABANA-TEBIBEL T., MELLOUK A., "Empirical QoE/QoS correlation model based on multiple parameters for VoD flows", *IEEE GlobeCom 2012, International Conference on Global Communications*, Anaheim, California, 2012.

[ATK 07] ATKINSON R., FANTO M., RIPv2 cryptographic authentication, Technical report, RFC 4822, 2007.

[AUE 95] AUER P., CESA-BIANCHI N., FREUND Y., *et al.*, "Gambling in a rigged casino: the adversarial multi-armed bandit problem", *36th Annual Symposium on Foundations of Computer Science*, IEEE, Milwaukee, WI, 1995.

[AUE 02] AUER P., CESA-BIANCHI N., FISCHER P., "Finite-time analysis of the multiarmed bandit problem", *Machine Learning*, vol. 47, no. 2, pp. 235–256, 2002.

[AUE 03] AUER P., CESA-BIANCHI N., FREUND Y., *et al.*, "The nonstochastic multiarmed bandit problem", *SIAM Journal on Computing*, vol. 32, no. 1, pp. 48–77, 2003.

[AUE 09] AUER P., JAKSCH T., ORTNER R., "Near-optimal regret bounds for reinforcement learning", *Journal of Machine Learning Research*, vol. 11, pp. 1563–1600, 2010

[AWE 02] AWEYA J., "Transmission control protocol", *Encyclopedia of Telecommunications*, Wiley Online Library, John Wiley & Sons, 2002.

[AWE 08] AWERBUCH B., KLEINBERG R., "Online linear optimization and adaptive routing", *Journal of Computer and System Sciences*, vol. 74, no. 1, pp. 97–114, 2008.

[BAR 96] BARTAL Y., "Probabilistic approximation of metric spaces and its algorithmic applications", *37th Annual Symposium on Foundations of Computer Science*, IEEE, Burlington, Vermont, pp. 184–193, 1996.

[BAT 10] BATTERAM H., DAMM G., MUKHOPADHYAY A., *et al.*, "Delivering quality of experience in multimedia networks", *Bell Labs Technical Journal*, vol. 15, no. 1, pp. 175–193, 2010.

[BEC 98] BECK M., MOORE T., "The Internet2 distributed storage infrastructure project: an architecture for internet content channels", *Computer Networks and ISDN Systems*, vol. 30, no. 22–23, pp. 2141–2148, 1998.

[BEL 59] BELLMAN R., KALABA R., "On adaptive control processes", *IRE Transactions on Automatic Control*, vol. 4, no. 2, pp. 1–9, 1959.

[BER 08] BERNARDO F., VUČEVIĆ N., UMBERT A., *et al.*, "Quality of experience evaluation under QoS-aware mobility mechanisms", *Proceedings of the 14th European Wireless Conference*, Prague, Czech Republic, September 2008.

[BHA 97] BHATTACHARJEE S., AMMAR M., ZEGURA E., *et al.*, "Application-layer anycasting", *Proceedings of the 16th Annual Joint Conference of the IEEE Computer and Communications (INFOCOM '97)*, Kobe, Japan, 1997.

[BIT 10a] BITAM S., BATOUCHE M., MELLOUK A., "QoSBeeManet: a new QoS multipath routing protocol for mobile ad-hoc networks", *GLOBECOM Workshops*, IEEE, Miami, Florida, 2010.

[BIT 10b] BITAM S., BATOUCHE M., TALBI E., "A survey on bee colony algorithms", *IEEE International Symposium on Parallel & Distributed Processing, Workshops and Phd Forum (IPDPSW)*, IEEE, Atlanta, GA, 2010.

[BOC 10] BOCCI M., BRYANT S., FROST D., *et al.*, "A framework for MPLS in transport networks", *IETF, RFC5921*, ISSN: 2070–1721, 2010.

[BOY 94] BOYAN J., LITTMAN M., "Packet routing in dynamically changing networks: a reinforcement learning approach", *Advances in Neural Information Processing Systems*, vol. 6, pp. 671–678, 1994.

[BRO 95] BROADBAND FORUM, Broadband Forum, available at http://www.broadband-forum.org/, 1995.

[BRO 10] BROOKS P., HESTNES B., "User measures of quality of experience: why being objective and quantitative is important", *IEEE, Network*, vol. 24, no. 2, pp. 8–13, 2010.

[BUB 08] BUBECK S., MUNOS R., STOLTZ G., *et al.*, "Online optimization in *X*-armed bandits", *Neural Information Processing Systems Conference (NIPS)*, Vancouver, Canada, 2008.

[CES 98] CESA-BIANCHI N., FISCHER P., "Finite-time regret bounds for the multiarmed bandit problem", *Proceedings of the 15th International Conference on Machine Learning*, Madison, WI, 1998.

[CHE 06] CHEN K., HUANG C., HUANG P., *et al.*, "Quantifying Skype user satisfaction", *ACM SIGCOMM Computer Communication Review*, vol. 36, pp. 399–410, 2006.

[CHE 07] CHEN T., "Network traffic modeling", *Handbook of Computer Networks*, John Wiley & Sons, pp. 326–339, 2007.

[CHE 10] CHELLOUCHE S., NEGRU D., BORCOCI E., *et al.*, "Anycast-based context-aware server selection strategy for VoD services", *GLOBECOM Workshops (GC Wkshps) 2010 IEEE*, Miami, Florida, 2010.

[CHI 11] CHIKKERUR S., SUNDARAM V., REISSLEIN M., *et al.*, "Objective video quality assessment methods: a classification, review, and performance comparison", *IEEE Transactions on Broadcasting*, vol. 57, no. 2, pp. 165–182, 2011.

[CIU 09] CIUBOTARU B., MUNTEAN G., "SAHA–a quality-oriented handover algorithm for multimedia content delivery to mobile users", *IEEE Transactions on Broadcasting*, vol. 55, no. 2, pp. 437–450, May 2009.

[COQ 07] COQUELIN P., MUNOS R., "Bandit algorithms for tree search", *Proceedings of the 23rd Conference on Uncertainty in Artificial Intelligence (UAI07)*, Vancouver, Canada, 2007.

[DEL 99] DELGADILLO K., Cisco distributed director, Cisco White Paper, Cisco, 1999.

[DIC 05] DI CARO G., DUCATELLE F., GAMBARDELLA L., "AntHocNet: an adaptive nature-inspired algorithm for routing in mobile ad hoc networks", *European Transactions on Telecommunications*, vol. 16, no. 5, pp. 443–455, 2005.

[DIG 96] DIGITAL ISLAND, Digital Island, available at http://www.digitalisland.com, 1996.

[DIL 02] DILLEY J., MAGGS B., PARIKH J., *et al.*, "Globally distributed content delivery", *IEEE, Internet Computing*, vol. 6, no. 5, pp. 50–58, 2002.

[DOR 06] DORIGO M., BIRATTARI M., STUTZLE T., "Ant colony optimization", *IEEE, Computational Intelligence Magazine*, vol. 1, no. 4, pp. 28–39, 2006.

[DRE 10] DRESSLER F., AKAN O., "Bio-inspired networking: from theory to practice", *Communications Magazine, IEEE*, vol. 48, no. 11, pp. 176–183, 2010.

[EDG 06] EDGECAST, "EdgeCast Networks", available at http://www.edgecast.com, 2006.

[ERT 09] ERTEL W., SCHNEIDER M., CUBEK R., *et al.*, "The teaching-box: a universal robot learning framework", *International Conference on Advanced Robotics (ICAR)*, IEEE, Munich, Germany, 2009.

[ESF 08] ESFAHANI A., ANALOUI M., "Widest K-shortest paths Q-routing: a new QoS routing algorithm in telecommunication networks", *International Conference on Computer Science and Software Engineering*, IEEE, Wuhan, China, 2008.

[EVE 02] EVEN-DAR E., MANNOR S., MANSOUR Y., "PAC bounds for multi-armed bandit and Markov decision processes", *Proceedings of the 15th Conference on Computational Learning Theory (COLT02)*, Sydney, Australia, 2002.

[FIE 10a] FIEDLER M., HLAVACS H., HACKBARTH K., *et al.*, "Quality of Experience", *Annals of Telecommunications*, vol. 65, no. 1, pp. 1–2, 2010.

[FIE 10b] FIEDLER M., HOSSFELD T., TRAN-GIA P., "A generic quantitative relationship between quality of experience and quality of service", *IEEE, Network*, vol. 24, no. 2, pp. 36–41, 2010.

[GAB 09] GABILLON V., Machine learning tools for online advertisement, Technical report, INRIA, Lille, France, 2009.

[GAD 97] GADDE S., RABINOVICH M., CHASE J., "Reduce, reuse, recycle: an approach to building large internet caches", *6th Workshop on Hot Topics in Operating Systems*, IEEE Cape Cod, MA, pp. 93–98, 1997.

[GAR 79] GARY M., JOHNSON D., "Computers and intractability: a guide to the theory of NP-completeness", *A Series of Books in the Mathematical Sciences*, W. H. FREEMAN and CO., VICTOR KLEE (ed.), 1979.

[GAR 93] GARCIA-LUNES-ACEVES J., "Loop-free routing using diffusing computations", *IEEE/ACM Transactions on Networking (TON)*, vol. 1, no. 1, pp. 130–141, 1993.

[GEL 04] GELENBE E., LENT R., "Power-aware ad hoc cognitive packet networks", *Ad Hoc Networks*, vol. 2, no. 3, pp. 205–216, 2004.

[GEL 06] GELLY S., WANG Y., MUNOS R., *et al.*, Modification of UCT with patterns in monte-carlo go, Technical report, INRIA, France, 2006.

[GEO 11] GEORGII H., *Gibbs Measures and Phase Transitions*, vol. 9, Walter de Gruyter, 2011.

[GOJ 03] GOJMERAC I., ZIEGLER T., RICCIATO F., *et al.*, "Adaptive multipath routing for dynamic traffic engineering", *Global Telecommunications Conference, 2003. GLOBECOM'03.* IEEE, San Francisco, CA, 2003.

[GOO 09] GOONG Y., YANG F., HUANG L., *et al.*, "Model-based approach to measuring quality of experience", *1st International Conference on Emerging Network Intelligence*, Sliema, Malta, 2009.

[GUE 99] GUERIN R., ORDA A., "QoS routing in networks with inaccurate information: theory and algorithms", *IEEE/ACM Transactions on Networking (TON)*, vol. 7, no. 3, pp. 350–364, 1999.

[GUO 02] GUO M., AMMAR M., ZEGURA E., "Selecting among replicated batching video-on-demand servers", *Proceedings of the 12th International Workshop on Network and Operating Systems Support for Digital Audio and Video*, ACM, Miami, FL, 2002.

[HAM 08] HAMAM A., EID M., EL SADDIK A., *et al.*, "A quality of experience model for haptic user interfaces", *Proceedings of the 2008 Ambi-Sys Workshop on Haptic User Interfaces in Ambient Media Systems*, ICST (Institute for Computer Sciences, Social-Informatics and Telecommunications Engineering), Quebec, Canada, 2008.

[HAN 09] HAN Y., KIM M., PARK H., "A novel server selection method to achieve delay-based fairness in the server palm", *IEEE Communications Letters*, vol. 13, no. 11, pp. 868–870, 2009.

[HAR 92] HARDWICK J., STOUT Q., *Bandit Strategies for Ethical Sequential Allocation*, Citeseer, 1992.

[HAS 10] HASSAN J., DAS S., HASSAN M., *et al.*, "Improving quality of experience for network services [Guest Editorial]", *IEEE Network*, vol. 24, no. 2, pp. 4–6, 2010.

[HE 08] HE J., REXFORD J., "Toward internet-wide multipath routing", *IEEE Network*, vol. 22, no. 2, pp. 16–21, 2008.

[HEI 08] HEINEMAN G., POLLICE G., SELKOW S., *Algorithms in a Nutshell*, O'Reilly Media, 2008.

[HEN 02] HENRY P., LUO H., "WiFi: what's next?", *IEEE Communications Magazine*, vol. 40, no. 12, pp. 66–72, 2002.

[HOC 04] HOCEINI S., *Techniques d'Apprentissage par Renforcement pour le Routage Adaptatif dans les Réseaux de Télécommunication à Trafic irrégulier*, PhD Thesis, University of Paris-Est Creteil, 2004.

[HOC 08] HOCEINI S., MELLOUK A., SMAIL B., "Average-bandwidth delay Q-routing adaptive algorithm", *IEEE International Conference on Communications*, ICC'08 IEEE Beijing, China, 2008.

[HUF 02] HUFFAKER B., FOMENKOV M., PLUMMER D., *et al.*, "k claffy distance metrics in the internet", *IEEE International Telecommunications Symposium*, Citeseer, Natal, Brazil, 2002.

[HUY 08] HUYNH-THU Q., GHANBARI M., "Scope of validity of PSNR in image/video quality assessment", *Electronics Letters*, vol. 44, no. 13, pp. 800–801, 2008.

[ISO 98] ISO/IEC, "Quality of service: framework", ISO/IEC 13236, 1998.

[ITU 65] ITU, "International Telecommunication Union", available at www.itu.int, 1865.

[JAK 10] JAKSCH T., ORTNER R., AUER P., "Near-optimal regret bounds for reinforcement learning", *The Journal of Machine Learning Research*, vol. 11, pp. 1563–1600, 2010.

[JAM 00] JAMIN S., JIN C., JIN Y., *et al.*, "On the placement of Internet instrumentation", *Proceedings of the 19th IEEE International Conference on Computer Communications (INFOCOM 2000)*, Tel Aviv, Israel, 2000.

[JAM 01] JAMIN S., JIN C., KURC A., *et al.*, "Constrained mirror placement on the Internet", *Proceedings of the 20th IEEE International Conference on Computer Communications (INFOCOM 2001)*, Anchorage, AK, 2001.

[KAA 01] KAARANEN H., AHTIAINEN A., LAITINEN L., *et al.*, *UMTS Networks*, Wiley Online Library, 2001.

[KAE 96] KAELBLING L.P., LITTMAN M.L., MOORE A.W., "Reinforcement learning: a survey", *Journal of AI Research*, vol. 4, pp. 237–285, 1996.

[KAR 99] KARGER D., SHERMAN A., BERKHEIMER A., *et al.*, "Web caching with consistent hashing", *Computer Networks*, vol. 31, no. 11–16, pp. 1203–1213, 1999.

[KAR 05] KARGIN V., "On the Chernoff bound for efficiency of quantum hypothesis testing", *The Annals of Statistics*, vol. 33, no. 2, pp. 959–976, 2005.

[KEP 03] KEPHART J., CHESS D., "The vision of autonomic computing", *Computer*, vol. 36, no. 1, pp. 41–50, 2003.

[KIL 02] KILKKI K., "Differentiated services for the Internet", *Interpretation*, vol. 1, pp. 00–99, 2002.

[KIM 08] KIM H.J., LEE D.H., LEE J.M., *et al.*, "The QoE evaluation method through the QoS-QoE correlation model", *4th International Conference on Networked Computing and Advanced Information Management*, Gyeongju, Korea, 2008.

[KIN 93] KINGMAN J., *Poisson Processes*, Wiley Online Library, 1993.

[KLE 08] KLEINBERG R., SLIVKINS A., UPFAL E., "Multi-armed bandits in metric spaces", *Proceedings of the 40th Annual ACM Symposium on Theory of Computing*, ACM Victoria, British Columbia, Canada, pp. 681–690, 2008.

[KOC 06] KOCSIS L., SZEPESVÁRI C., "Bandit based monte-carlo planning", *Machine Learning: ECML 2006*, vol. 4212, pp. 282–293, 2006.

[KOH 06] KOHLER E., HANDLEY M., FLOYD S., "Designing DCCP: congestion control without reliability", *ACM SIGCOMM Computer Communication Review*, vol. 36, no. 4, pp. 27–38, 2006.

[KOO 06] KOONEN T., "Fiber to the home/fiber to the premises: what, where, and when?", *Proceedings of the IEEE*, vol. 94, no. 5, pp. 911–934, 2006.

[KOU 08] KOULOURIOTIS D., XANTHOPOULOS A., "Reinforcement learning and evolutionary algorithms for non-stationary multi-armed bandit problems", *Applied Mathematics and Computation*, vol. 196, no. 2, pp. 913–922, 2008.

[KRI 00] KRISHNAN P., RAZ D., SHAVITT Y., "The cache location problem", *IEEE/ACM Transactions on Networking (TON)*, vol. 8, no. 5, pp. 568–582, 2000.

[KUR 05] KUROSE J., ROSS K., *Computer Networking*, Pearson/Addison Wesley, 2005.

[LAI 85] LAI T., ROBBINS H., "Asymptotically efficient adaptive allocation rules", *Advances in Applied Mathematics*, vol. 6, no. 1, pp. 4–22, 1985.

[LEM 10] LEMMA E., HUSSAIN S., ANJELO W., Performance comparison of EIGRP/IS-IS and OSPF/IS-IS, Master's Thesis, Blekinge Tekniska Hogskolan, Sweden, 2010.

[LI 08] LI J., Agent organization in the knowledge plane, PhD Thesis, Cambridge, MA, 2008.

[LIM 01] LIMELIGHT, Limelight, available at http://www.limelight.com, 2001.

[LIT 94] LITTMAN M., "Markov games as a framework for multi-agent reinforcement learning", *Proceedings of the 1th International Conference on Machine Learning*, New Brunswick, NJ, 1994.

[LOP 11] LOPES GOMES R., MOREIRA JUNIOR W., CERQUEIRA E., *et al.*, "Using fuzzy link cost and dynamic choice of link quality metrics to achieve QoS and QoE in wireless mesh networks", *Journal of Network and Computer Applications*, vol. 34, no. 2, pp. 506–516, 2011.

[LUC 59] LUCE R., *Individual Choice Behavior*, John Wiley, 1959.

[MA 97] MA Q., STEENKISTE P., "On path selection for traffic with bandwidth guarantees", *Proceedings of the International Conference on Network Protocols*, IEEE, Atlanta, GA, 1997.

[MAJ 09] MAJD G., CESAR V., ADLEN K., "An adaptive mechanism for multipath video streaming over video distribution network (VDN)", *Proceedings of the 1st International Conference on Advances in Multimedia (MMEDIA 2009)*, Colmar, France, 2009.

[MAN 04] MANNOR S., TSITSIKLIS J., "The sample complexity of exploration in the multi-armed bandit problem", *The Journal of Machine Learning Research*, vol. 5, pp. 623–648, 2004.

[MAO 02] MAO Z., CRANOR C., DOUGLIS F., *et al.*, "A precise and efficient evaluation of the proximity between web clients and their local DNS servers", *USENIX Annual Technical Conference*, Monterey, Canada, June 2002.

[MAR 91] MARYLAND ROUTING SIMULATOR, "Maryland routing simulator", available at http://www.cs.umd.edu/projects/netcalliper/software.html, 1991.

[MAR 07a] MARKAKI O., CHARILAS D., NIKITOPOULOS D., "Enhancing quality of experience in next generation networks through network selection mechanisms", *18th International Symposium on Personal, Indoor and Mobile Radio Communications, 2007. PIMRC 2007*, IEEE Athens, Greece, pp. 1–5, 2007.

[MAR 07b] MARKOVIĆ G., TEODOROVIĆ D., AĆIMOVIĆ-RASPOPOVIĆ V., "Routing and wavelength assignment in all-optical networks based on the bee colony optimization", *AI Communications*, vol. 20, no. 4, pp. 273–285, 2007.

[MAR 10] MARTINEZ-YELMO I., SEOANE I., GUERRERO C., "Fair Quality of Experience (QoE) measurements related with networking technologies", *Proceedings of the 8th International Conference on Wired/Wireless Internet Communications*, Lulea, Sweden, 2010.

[MEL 94] MELLOUK A., Etude des sytèmes modulaires neuro-prédictifs: application à la reconnaissance automatique de la parole continue, PhD Thesis, University Paris-Sud, Orsay, France, 1994.

[MEL 05] MELLOUK A., HOCEINI S., "A reinforcement learning approach for QoS based routing packets in integrated service web based systems", *Lecture Notes in Artificial Intelligence*, vol. 3528, pp. 299–305, 2005.

[MEL 09] MELLOUK A., HOCEINI S., ZEADALLY S., "Design and performance analysis of an inductive QoS routing algorithm", *Computer Communications*, Elsevier Ed., vol. 32, pp. 1371–1376, 2009.

[MEL 11a] MELLOUK A., *Advances in Reinforcement Learning*, InTechWeb Education and Publishing Ed., Open Acess, 2011.

[MEL 11b] MELLOUK A., HOCEINI S., ZEADALLY S., "A bio-inspired quality of service (QoS) routing algorithm", *IEEE Communications Letters*, vol. 15, no. 9, pp. 1016–1018, 2011.

[MEL 13] MELLOUK A., HOCEINI S., ZEADALLY S., "A state-dependent time evolving multi-constraint routing algorithm", *ACM Transactions on Autonomous and Adaptive Systems (TAAS)*, ACM, vol. 8, no. 1, pp. 1–21, 2013.

[MER 09] MERSEREAU A., RUSMEVICHIENTONG P., TSITSIKLIS J., "A structured multiarmed bandit problem and the greedy policy", *IEEE Transactions on Automatic Control*, vol. 54, no. 12, pp. 2787–2802, 2009.

[MIC 79] MICHAEL R., JOHNSON D., *Computers and Intractability: A Guide to the Theory of NP-Completeness*, WH Freeman & Co., San Francisco, 1979.

[MOE 08] MOEBS S., "A learner, is a learner, is a user, is a customer – QoS-based experience-aware adaptation", *16th ACM International Conference on Multimedia*, Vancouver, Canada, 2008.

[MOY 08] MOY J., *OSPF Complete Implementation*, Addison-Wesley Professional, 2008.

[NET 89] NETWORK SIMULATOR, "The network simulator", available at http://www.isi.edu/nsnam/ns/, 1989.

[NLA 06] NLANR, "National laboratory for applied network research", available at http://www.nlanr.net/, 2006.

[NUA 07] NUAYMI L., *WiMAX: Technology for Broadband Wireless Access*, Wiley, 2007.

[OKA 09] OKAMURA H., DOHI T., TRIVEDI K., "Markovian arrival process parameter estimation with group data", *IEEE/ACM Transactions on Networking (TON)*, vol. 17, no. 4, pp. 1326–1339, 2009.

[PAN 03] PAN J., HOU Y., LI B., "An overview of DNS-based server selections in content distribution networks", *Computer Networks*, vol. 43, no. 6, pp. 695–711, 2003.

[PAT 08] PATHAN M., BUYYA R., "A taxonomy of CDNs", *Content Delivery Networks*, Springer Ed., pp. 33–77, 2008.

[PAT 12] PATEL A., YADAV M., "Optimization of routing algorithm using fuzzy neural network", *Asian Journal of Computer Science and Information Technology*, vol. 2, no. 7, pp. 180-183, 2012.

[PCM 94] PC MAG, "PC Magazine Digital Edition", available at http://www.pcmag.com/, 1994.

[PEN 04] PENG G., CDN: Content distribution network, Technical report, Experimental Computer Systems Lab, Department of Computer Science, State University of New York, NY, 2004.

[PIC 97] PICARD R., *Affective Computing*, The MIT Press, Cambridge, MA, vol. 167, p. 170, 1997.

[PIE 06] PIERRE G., VAN STEEN M., "Globule: a collaborative content delivery network", *IEEE Communications Magazine*, vol. 44, no. 8, pp. 127–133, 2006.

[QIU 01] QIU L., PADMANABHAN V., VOELKER G., "On the placement of web server replicas", *Proceedings of the 20th IEEE International Conference on Computer Communications (INFOCOM 2001)*, Anchorage, AK, 2001.

[RAD 08] RADLINSKI F., KLEINBERG R., JOACHIMS T., "Learning diverse rankings with multi-armed bandits", *Proceedings of the 25th International Conference on Machine Learning*, ACM, New York, NY, 2008.

[RAM 00] RAMALHO M., "Intra-and inter-domain multicast routing protocols: a survey and taxonomy", *IEEE Communications Surveys & Tutorials*, vol. 3, no. 1, pp. 2–25, 2000.

[REC 94] RECOMMENDATION, "ITU E.800: terms and definitions related to quality of service and network performance including dependability", ITU Telecom, 1994.

[REC 99] RECOMMENDATION, "ITU-T P.910: subjective video quality assessment methods for multimedia applications", ITU Telecom, 1999.

[REC 01a] RECOMMENDATION, "ITU-T G.1010: end-user multimedia QoS categories", ITU Telecom, 2001.

[REC 01b] RECOMMENDATION, "ITU-T P.862: perceptual evaluation of speech quality (PESQ): an objective method for end-to-end speech quality assessment of narrow-band telephone networks and speech codecs ", ITU Telecom, 2001.

[REC 02] RECOMMENDATION, "ITU-R BT.500-11: methodology for the subjective assessment of the quality of television pictures", ITU Telecom, 2002.

[REC 04a] RECOMMENDATION, "ITU-R BT.1683: objective perceptual video quality measurement techniques for standard definition digital broadcast television in the presence of a full reference", ITU Telecom, STANDARD published by International Telecommunications Union, 2004.

[REC 04b] RECOMMENDATION, "ITU-T J.144: objective perceptual video quality measurement techniques for digital cable television in the presence of a full reference", ITU Telecom, 2004.

[REC 05] RECOMMENDATION, "ITU-T G.1030: estimating end-to-end performance in IP networks for data applications", ITU Telecom, 2005.

[REC 06] RECOMMENDATION, "ITU P.801: mean opinion score (MOS) terminology", ITU Telecom, 2006.

[REC 07] RECOMMENDATION, "ITU-T G.1070: opinion model for video-telephony applications", ITU Telecom, 2007.

[REC 08a] RECOMMENDATION, "ITU-T J.246: perceptual visual quality measurement techniques for multimedia services over digital cable television networks in the presence of a reduced bandwidth reference", ITU Telecom, 2008.

[REC 08b] RECOMMENDATION, "ITU-T J.247 objective perceptual multimedia video quality measurement in the presence of a full reference", ITU Telecom, 2008.

[REC 10] RECOMMENDATION, "ITU-T J.249: perceptual video quality measurement techniques for digital cable television in the presence of a reduced reference", ITU Telecom, 2010.

[REC 11a] RECOMMENDATION, "ITU-T G.1000 qualité de service des communications: cadre et définitions", ITU Telecom, 2011.

[REC 11b] RECOMMENDATION, "ITU-T G.107: the E-model: a computational model for use in transmission planning", ITU Telecom, 2011.

[REC 11c] RECOMMENDATION, "ITU-T J.342: objective multimedia video quality measurement of HDTV for digital cable television in the presence of a reduced reference signal", ITU Telecom, 2011.

[RÉT 09] RÉTVÁRI G., NÉMETH F., CHAPARADZA R., *et al.*, "Ospf for implementing self-adaptive routing in autonomic networks: a case study", *Modelling Autonomic Communications Environments*, Springer Berlin Heidelberg, Lecture Notes in Computer Science, vol. 5844, pp. 72–85, 2009.

[RIN 09] RINGSERVER, "RingServer project", available at http://www.ring.gr.jp/, 2009.

[RIX 02] RIX A., HOLLIER M., HEKSTRA A., *et al.*, "Perceptual evaluation of speech quality (PESQ): the new ITU standard for end-to-end speech quality assessment part I-time-delay compensation", *Journal of the Audio Engineering Society*, vol. 50, no. 10, pp. 755–764, 2002.

[ROB 52] ROBBINS H., "Some aspects of the sequential design of experiments", *Bulletin of the American Mathematical Society*, vol. 58, no. 5, pp. 527–535, 1952.

[ROH 00] ROHALY A., LIBERT J., CORRIVEAU P., *et al.*, "Final report from the video quality experts group on the validation of objective models of video quality assessment", *ITU-T Standards Contribution COM*, pp. 9–80, 2000.

[ROU 98] ROUSSKOV A., WESSELS D., "Cache digests", *Computer Networks and ISDN Systems*, vol. 30, no. 22–23, pp. 2155–2168, 1998.

[RUB 05] RUBINO G., "Quantifying the quality of audio and video transmissions over the internet: the PSQA approach", *Design and Operations of Communication Networks: A Review of Wired and Wireless Modeling and Management Challenges*, Imperial College Press, London, UK, pp. 233–248, 2005.

[RUS 10] RUSMEVICHIENTONG P., TSITSIKLIS J., "Linearly parameterized bandits", *Mathematics of Operations Research*, vol. 35, no. 2, pp. 395–411, 2010.

[SCH 09] SCHULER W., BASTOS-FILHO C., OLIVEIRA A., "A novel hybrid training method for hopfield neural networks applied to routing in communications networks", *International Journal of Hybrid Intelligent Systems*, vol. 6, no. 1, pp. 27–39, 2009.

[SCH 10] SCHAUL T., BAYER J., WIERSTRA D., *et al.*, "PyBrain", *The Journal of Machine Learning Research*, vol. 11, pp. 743–746, 2010.

[SHA 01] SHAIKH A., TEWARI R., AGRAWAL M., "On the effectiveness of DNS-based server selection", *Proceedings of the 20th IEEE International Conference on Computer Communications (INFOCOM 2001)*, Anchorage, AK, 2001.

[SHA 10a] SHAIKH J., FIEDLER M., COLLANGE D., "Quality of experience from user and network perspectives", *Annals of Telecommunications*, vol. 65, no. 1, pp. 47–57, 2010.

[SHA 10b] SHARMA M., KAUR H., *et al.*, "Performance evaluation of hybrid network using RIP & IGRP for different applications", *High-Capacity Optical Networks and Enabling Technologies, HONET 2010*, IEEE, Cairo, Egypt, 2010.

[SOL 06] SOLDANI D., LI M., CUNY R., *QoS and QoE Management in UMTS Cellular Systems*, Wiley Online Library, 2006.

[SOU 12] SOUIHI S., HOCEINI S., MELLOUK A., *et al.*, "A hierarchical and multi-criteria knowledge dissemination in autonomic networks", *IEEE GlobeCom 2012, International Conference on Global Communications*, IEEE, Atlanta, GA, 2012.

[STA 99] STARR T., CIOFFI J., SILVERMAN P., *Understanding Digital Subscriber Line Technology*, Prentice Hall PTR, 1999.

[STE 01] STEWART R., METZ C., "SCTP: new transport protocol for TCP/IP", *IEEE Internet Computing*, vol. 5, no. 6, pp. 64–69, 2001.

[SUT 98] SUTTON R.S., BARTO A.G., *Reinforcement Learning: An Introduction*, MIT Press, Cambridge, 1998.

[SZY 03] SZYMANIAK M., PIERRE G., VAN STEEN M., "NetAirt: a flexible redirection system for apache", *International Conference WWW/Internet (ICWI)*, Algarve, Portugal, 2003.

[TAK 08] TAKAHASHI A., HANDS D., BARRIAC V., "Standardization activities in the ITU for a QoE assessment of IPTV", *IEEE Communications Magazine*, vol. 46, no. 2, pp. 78–84, 2008.

[TAN 09] TANNER B., WHITE A., "RL-Glue: language-independent software for reinforcement-learning experiments", *The Journal of Machine Learning Research*, vol. 10, pp. 2133–2136, 2009.

[TEO 05] TEODOROVIĆ D., DELL'ORCO M., "Bee colony optimization – a cooperative learning approach to complex transportation problems", *Advanced OR and AI Methods in Transportation, Proceedings of 16th Mini–EURO Conference and 10th Meeting of EWGT (13–16 September 2005)*, Publishing House of the Polish Operational and System Research, Poznan, Poland, 2005.

[TMF 88] TM FORUM, available at http://www.tmforum.org/, 1988.

[TRA 10] TRAN H.A., MELLOUK A., HOCEINI S., *et al.*, QoE-based concepts and models for end-to-end network services based on user perception, Technical report, Laboratory LISSI, University Paris-Est Creteil, 2010.

[TRA 13a] TRAN H.A., MELLOUK A., HOCEINI S., *et al.*, "A dynamic QoE routing system for user-centric applications", *Transactions on Emerging Telecommunications Technologies*, vol. 24, no. 3, pp. 266–279, 2013.

[TRA 13b] TRAN H.A., MELLOUK A., PEREZ J., *et al.*, "QoE-based server selection for content distribution networks", *IEEE Transactions on Computers*, DOI:http://doi.ieeecomputersociety.org/10.1109/TC.2013.33, to appear, 2013.

[TRI 10] TRINH T., SONKOLY B., MOLNÁR S., "On the fairness characteristics of Fast TCP", *Annals of Telecommunications*, vol. 65, no. 1, pp. 73–85, 2010.

[VAL 98] VALLOPILLIL V., ROSS K., "Internet draft: cache array routing protocol v1. 0", available at http://icp.ircache.net/carp.txt, 1998.

[VER 05] VERMOREL J., MOHRI M., "Multi-armed bandit algorithms and empirical evaluation", *Proceedings of the 16th European Conference on Machine Learning (ECML 2005)*, Porto, Portugal, 2005.

[VLE 08] VLEESCHAUWER B.D., TURCK F.D., DHOEDT B., *et al.*, "End-to-end QoE optimization through overlay network deployment", *International Conference on Information Networking*, Busan, Korea, 2008.

[WAN 95] WANG Z., CROWCROFT J., "Bandwidth-delay based routing algorithms", *Global Telecommunications Conference, GLOBECOM'95*, IEEE, Singapore, 1995.

[WAN 96] WANG Z., CROWCROFT J., "Quality-of-service routing for supporting multimedia applications", *IEEE Journal on Selected Areas in Communications*, vol. 14, no. 7, pp. 1228–1234, 1996.

[WAN 04] WANG Z., BOVIK A., SHEIKH H., *et al.*, "Image quality assessment: from error visibility to structural similarity", *IEEE Transactions on Image Processing*, vol. 13, no. 4, pp. 600–612, 2004.

[WAN 08] WANG Y., MUNOS R., AUDIBERT J., "Algorithms for infinitely many-armed bandits", *Proceeding of Advances in Neural Information Processing Systems*, Vancouver, Canada, 2008.

[WAT 89] WATKINS H., Learning from delayed rewards, PhD Thesis, Cambridge University, 1989.

[WAT 92a] WATKINS, DAYLAN, "Technical note: Q-learning", *Machine Learning*, vol. 8, no. 3, pp. 279–292, 1992.

[WAT 92b] WATKINS C., DAYAN P., "Q-learning", *Machine Learning*, vol. 8, no. 3, pp. 279–292, 1992.

[WAX 93] WAXMAN B., "Performance evaluation of multipoint routing algorithms", *Proceedings of the 12th IEEE International Conference on Computer Communications (NFOCOM'93)*, San Francisco, CA, 1993.

[WEL 06] WELCH J., CLARK J., "RFC4445 – a proposed media delivery index (MDI)", IETF, RFC4445, 2006.

[WEN 10] WENDELL P., JIANG J., FREEDMAN M., *et al.*, "Donar: decentralized server selection for cloud services", *ACM SIGCOMM Computer Communication Review*, ACM, vol. 40, pp. 231–242, 2010.

[WES 97] WESSELS D., CLAFFY K., Application of internet cache protocol (ICP), version 2, 1997.

[WIJ 10] WIJNANTS M., LAMOTTE W., DE VLEESCHAUWER B., *et al.*, "Optimizing user quality of experience through overlay routing, bandwith management and dynamic transcoding", *International Journal of Adaptive, Resilient and Automatic Systems*, vol. 1, no. 2, p. 64, 2010.

[WIK 01] WIKIPEDIA, "Wikipedia, the free encyclopedia", available at http://wikipedia.org, 2001.

[YAM 07] YAMADA H., FUKUMOTO N., ISOMURA M., *et al.*, "A QoE based service control scheme for RACF in IP-based FMC networks", *E- Commerce Technology and the 4th IEEE International Conference on Enterprise Computing*, Tokyo, Japan, 2007.

[YIN 10] YIN H., LIU X., MIN G., *et al.*, "Content delivery networks: a bridge between emerging applications and future IP networks", *IEEE Network*, vol. 24, no. 4, pp. 52–56, 2010.

[YOS 97] YOSHIKAWA C., CHUN B., EASTHAM P., *et al.*, "Using smart clients to build scalable services", *Proceedings of the 1997 USENIX Technical Conference*, Anaheim, California, 1997.

[ZHA 07] ZHANG J., HU H., WANG T., *et al.*, "QOSPF-improved routing algorithm with bandwidth and delay metrics", *Beijing Youdian Daxue Xuebao/Journal of Beijing University of Posts and Telecommunication*, vol. 30, no. 4, pp. 134–138, 2007.

[ZHA 11] ZHANG J., ANSARI N., "On assuring end-to-end QoE in next generation networks: challenges and a possible solution", *IEEE Communications Magazine*, vol. 49, no. 7, pp. 185–191, 2011.

[ZHO 08] ZHOU X., DREIBHOLZ T., RATHGEB E., "A new server selection strategy for reliable server pooling in widely distributed environments", *2nd International Conference on the Digital Society*, IEEE Sainte Luce, Martinique, 2008.

Index

T, U, V

Printed and bound by CPI Group (UK) Ltd, Croydon, CR0 4YY

13/09/2026

14961081-0008